PIPE ECONOMICS

CONTENTS

ACKNOWLEDGMENTS

This book has taken shape over many years, and it would not exist without the intellectual and financial support of many individuals and institutions. Generous funding for its research and writing was provided by the Richard Hofstadter Fellowship from Columbia University, the Mellon Fellowship from the American Council of Learned Societies, and a postdoctoral fellowship hosted by the Max Planck Institute for the History of Science—Department III: Artefacts, Action, and Knowledge. I am grateful for the assistance and guidance of the staff and archivists at the Egyptian National Library and Archives, the Egyptian Survey Authority, the British National Archives, the Institution of Civil Engineers in London, the Rare Books and Special Collections Library of the American University in Cairo, and CEDEJ (Centre d'Études et de Documentation Économiques, Juridiques et Sociales) in Cairo. I owe special thanks to Adel Hasan Zaki—then head of the Projects Sector at the Greater Cairo Sanitary Drainage Company—for approving my visit to the original sites of Cairo's colonial-era drainage system.

I am grateful for the academic guidance and generous feedback that I received from many mentors and advisors. The origins of this book go back to a meeting with Khaled Fahmy in New York in 2008, where he shared with me his research on the "olfactory history" of Cairo. This eye-opening conversation inspired me to think about how public health history can inform other fields of historical writing. This book would not have been possible without our conversations since then. I first came to the idea of writing a history of Cairo through its pipes and sewers in late 2010, in the early days of the Arab Spring. At the time, I was reading Mike Davis's *City of Quartz: Excavating the Future in Los Angeles* with Robert Vitalis, under whose advice I became interested in economic and business history. Without his sharp insights, I may have never thought to study the economics of water management or to look seriously at the history of the Cairo Water Company. Meanwhile, Eve Troutt Powell's guidance was key in helping me think like a historian—having been mostly trained in philosophy and Middle East studies departments until then—and in piquing my interest in the history of Egypt during the British colonial period. My conversations with Marwa Elshaky were formative in shaping my interest in the history of science and technology. Without her gentle guidance and untiring support, I might never have considered writing a book about how large-scale technology has shaped modern society. Timothy Mitchell has been a source of inspiration throughout the research and writing of this book in so many ways, not the least of which was his advice to look into the field of railway economics. I was also fortunate to work with two historians of New York—Elizabeth Blackmar and Gergely Baics—from whom I learned so much about analyzing urban economies.

Throughout this long research and writing journey, more friends, colleagues, and even total strangers have helped, inspired, and given me feedback than I can possibly list in these pages. Thank you to Begum Adalet, Ali Al-Adawi, Iman Allam, Nader Andrawos, Mohammed Rafi Arefin, Jon Argaman, Wael Ashry, On Barak, Beth Baron, Salma Belal, Ahmad al-Bindary, Edna Bonhomme, Shady Lewis Botros, Omar Cheta, Deborah Coen, Jeremy Dell, Jennifer Derr, Hossam Elouan, Mohamed Elshahed, Amr Essam, Mohammed Ezzeldin, Wael Abd El-Fattah, Yaqin Fouad, Mohamed Gad, Aimee Genell, Noha El-Ghalban, Pascale Ghazaleh, Omar Ghazzi, Angela Giordani, Anthony Greco, Salvador El-Hady, Muhammad al-Hajj, Fouad Halbouni, Jens Hanssen, Melanie Henry, Murad Idris, Aaron Jakes, Ceyda Karamursel, Sarah El-Kazaz, Amr Khairy, Omnia Khalil, Malak Labib, Owain Lawson, Mai al-Maghrabi, Mariam Mekiwi, Pascal Ménoret, Sara Mourad, Mohammad Naeem, Ahmad Naji, Aya Nassar, Anne Norton, Tamar Novick, Lisa Onaga,

Daniel Payan, Émilie Pasquier, Seif El Rashidi, David Rosner, Lucie Ryzova, Ahmad Fawzi Saleh, Mahmud Salem, Dagmar Schäfer, Sherene Seikaly, Salma Shamel, Yahia Shawkat, Ahmad Shokr, Kavita Sivaramakrishnan, Pamela Smith, Dale Stahl, Said Abu Talib, Abdel-Rahman al-Taliawi, Dalia Wahdan, Sherif Younis, Karin Zachmann, Maged Zaher, and Adrien Zakar. I am especially grateful to Mai Taha, who has heard more about water and sewers than any nonspecialist likely should over several lifetimes.

My family has supported me throughout this journey. I have had the fortune to discuss parts of this book with my father, Fakhry Ismail, a mechanical engineer by training, which helped me decipher what water and sanitary engineers were doing in Cairo. I discussed the more theoretical aspects of Egypt's political economy with my mother, Mona Sadiq, an education specialist. In the course of writing this book, I rediscovered the writings of my maternal grandfather, Ahmad Sadiq Sa'd—a Marxist thinker, a leader in the 1940s communist movement, and an electrical engineer by training. It is to his memory and to my son, Zayd, that I dedicate this book.

NOTE ON TRANSLITERATION

I have followed the International Journal of Middle East Studies system for Arabic transliteration, simplified by omitting diacritical marks except for the hamza (') and 'ayn ('). Common spellings for place names (Cairo, Alexandria, Mecca) and personal names (Mehmed Ali, Gamal Abdel Nasser) have been retained. Additionally, I have preserved Arab authors' own transliterations of their names in publications in European languages. All translations are my own unless otherwise noted.

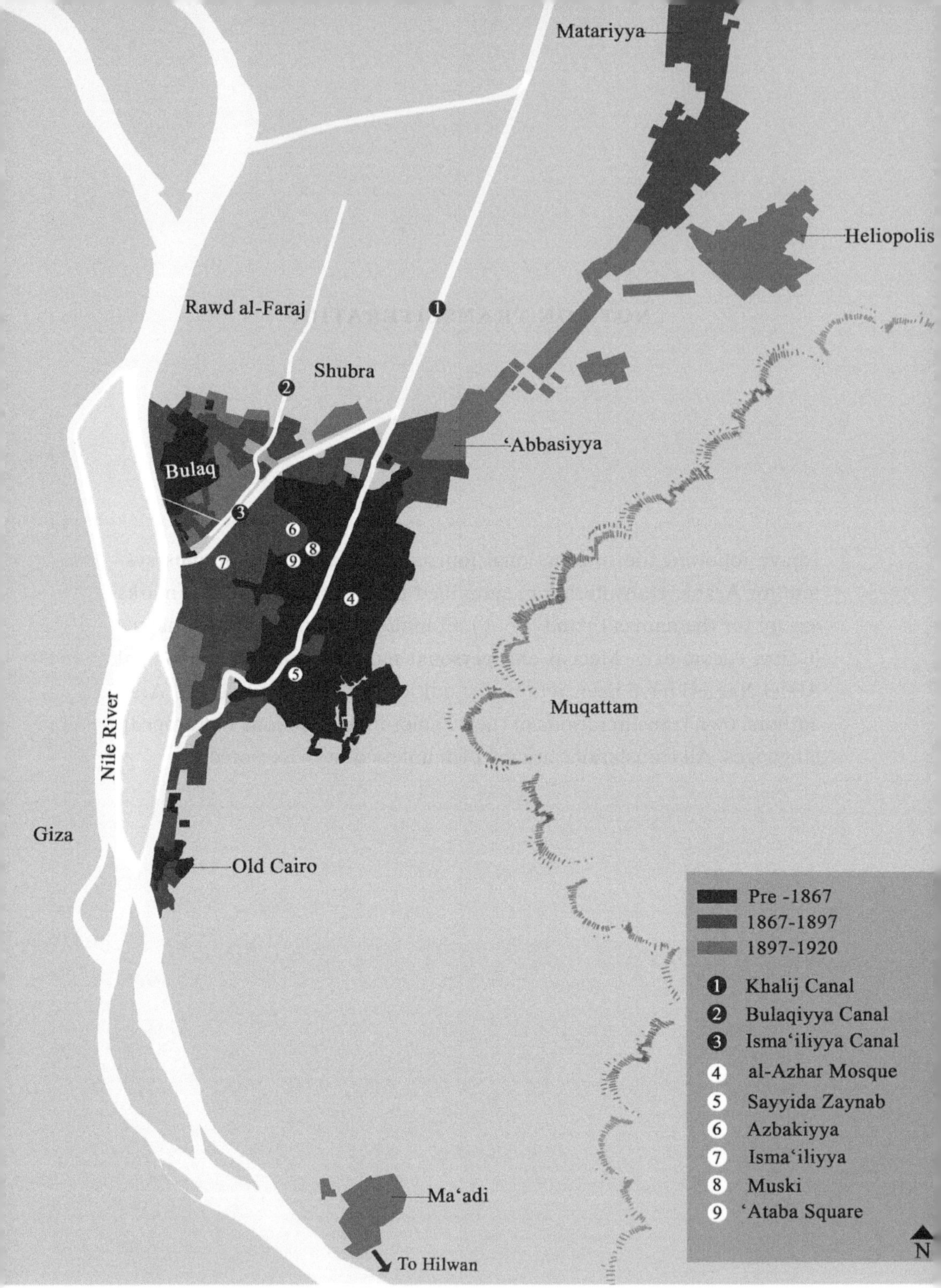

FIGURE 1. Orientation map. Illustration by Amr Essam.

INTRODUCTION

Piped Infrastructure and the Modern City

Pipes and sewers have long occupied a unique place in the urban imagination. Channeling both an essential element of life and the wastes of metabolic processes, they are often envisioned as the city's "intestines"—a visceral domain hidden from view, yet vital for survival of the urban "organism." Due to their visceral nature, they have also been perceived as conduits for the primal, archaic, and irrational elements of society. At the same time, water and sewer pipelines have been celebrated as marvels of engineering—testaments to how science and technology can be mobilized to meet the goals of convenience and health. In this interpretation, the industrialized underground represents a daring attempt to meticulously reorganize and improve the urban "machine" above.[1] Victor Hugo's epic novel *Les Misérables* offers a classic example of this fascination with subterranean spaces. Hugo devoted an entire chapter to the Parisian sewers—"The Bowels of Leviathan"—where the protagonist finds refuge during the uprising of June 1832. Digressing from the storyline, Hugo ruminates on the multiple meanings of sewers: as the city's underbelly, an artifact of modern engineering, and a site of regeneration and redemption. Ultimately, the sewers appear as a metaphor for how society manages—or hides—its waste, moral and material alike. Hence "the history of humankind is reflected in the history of sewers," he declared.[2]

This book seeks to write a new history of Cairo through its pipes and drains. It shows that the introduction of water infrastructures not only reconfigured the relationship between Egyptians and the Nile, but also had significant consequences for urban planning practices, public health regimes, housing and construction patterns, urban services and their underlying economies, and even national politics. Furthermore, it argues that these changes were paradigmatic of the global experience of cities when modern regimes of water management were first introduced. By putting the Egyptian capital in a broader context, *Pipe Economics* seeks to distinguish between the common and the particular aspects of this history, and to explore both on their own terms.

Born out of the industrial revolution, the global rollout of mechanized water supply and drainage networks took place over a long period spanning the nineteenth and twentieth centuries. Typically expensive and logistically demanding, these infrastructures were initially justified on public health grounds. By preventing the spread of waterborne diseases such as cholera, they removed the biological constraints on population growth and contributed to the unprecedented rise of urbanization in the modern era. However, their significance rapidly extended beyond sustaining healthy cities. Piped water and drains became indispensable conveniences, embodying the truism that the luxury of yesterday is the necessity of today. Wherever they extended, they were perceived as emblems of global modernity. By automating old methods of water delivery and waste disposal, which once relied on manual labor, these systems fundamentally changed not only how cities were built but also how people lived, consumed, cared for their bodily hygiene, and understood what it meant to be modern.

One of the core contentions of this book is that the transformative effects of water infrastructures on modern societies have not been fully recognized beyond a small circle of specialists. Open any book on the history of railways—a comparable nineteenth-century innovation in large-scale technology—and you are likely to find a description of a transportation revolution.[3] Indeed, railways drastically sped up the transport of people and goods across vast distances. By replacing horse-drawn carriages with steam-powered locomotives and metal rails, they produced far-reaching economic and political consequences and altered the very perception of space and time. Similarly, histories of the telegraph typically underscore its revolutionary impact on communication.[4] By radically increasing the speed of information from days, weeks, or even months to minutes or hours, telegraphs enabled real-time communication for the first time, with widespread effects on global business, the dissemination of news, and military and political control. But for reasons that will be discussed below, the introduction of water systems is usually not framed in similarly dramatic terms.

Pipe Economics seeks to challenge this imbalance, showing that the impact of piped networks was in fact comparable to that of railways and telegraphs. If the latter revolutionized modes of transportation and communication, water infrastructures revolutionized the very basic act of using water. By transforming key aspects of human dependence on water resources, the global rollout of these networks marked the end of a hydrological order based on manual labor and collective modes of water service and the beginning of another based on mechanization and private water consumption. In turn, this global revolution in water consumption had far-reaching consequences for everyday life, as well as urban geography, economy, and politics.

While positioning Cairo as a platform for exploring this near-universal story, this book also reflects on what was peculiar about how this transformation played out there. Cairo offers an important contrast to European and North American cities in that the installation of water systems was conditioned by imperialism from the very beginning. Similar to other cities with a colonial past, Western expertise and capital played a significant role in building Cairo's water infrastructures, which distinctively shaped their political and economic trajectory. The difference is most striking when considering the political economy of water management. Water provisioning in European and North American cities became subject to a host of government controls by the early twentieth century, including quality and price regulations, municipalization, and nationalization of private water companies.[5] In Cairo, by contrast, it was marked by the outsized power of a single foreign corporation—the

Société Anonyme des Eaux du Caire (Cairo Water Company)—formed in 1865 with predominantly French capital. Over the following decades, the company exercised monopoly rights, tightened its grip on "consumers," and generated huge profits. In 1957, its assets were seized by order of President Gamal Abdel Nasser, who had nationalized the Suez Canal a year earlier, igniting a major political and military conflict with the West. By tracing the history of this company from beginning to end, this book retells the history of modern Cairo pipe by pipe.

The Water Consumption Revolution in Global Perspective

Water supply and drainage networks have been as consequential to the making of global modernity as railways, telegraphs, telephones, the internal combustion engine, and industrialized steel production. But for various reasons, they did not receive nearly the same amount of attention. Whereas the transformative effects of the latter have been well explored and documented, those caused by water and drainage systems have been relatively less studied and understood. One of the main reasons for this oversight is the slow pace of change generated by them. Whereas the revolutionary impacts of railways were evident within a decade or two, the full effects of water infrastructures were relatively slow, taking several decades to materialize. This has made it relatively harder to detect their transformative character. However, slowly but surely, networks of water supply and wastewater removal introduced a new environmental regime of water access and consumption.

In quantitative terms, they have led to a considerable increase in water usage on average. Water distribution was historically dependent on the manual power of water carriers and draft animals, as well as preindustrial technologies such as water wheels and gravity-based aqueducts. This limited the volume of water that could be transported and the heights to which it could be raised. Mechanizing the water supply removed these limits, making far more water—and from sources further away or deeper underground—available at cities' disposal. Meanwhile, preindustrial cities commonly disposed wastewater in percolating cesspits or nearby rivers, canals, and lakes, which placed constraints on how much water could be used without causing nuisance or contaminating fresh water sources. With mechanized drainage, these limits were lifted. Before the introduction of these networks, it is difficult to imagine water consumption beyond 10–50 liters per capita per day, notwithstanding geographic differences and excluding outliers such as Ancient Rome with its colossal aqueducts.[6] In contemporary cities, the figure jumps to 100–400 liters, with notable regional variations.[7]

In qualitative terms, water systems facilitated a shift from public to private water usage. Traditionally, cities relied on shared facilities of water access—including communal baths, laundries, fountains, and kitchens—which commonly received their water supply in bulk. Meanwhile, a typical pre-modern household had to haul water from a well, stream, or public fountain. As a result, domestic water usage accounted for only a small fraction of total consumption. With the introduction of mechanized water supply and drainage, the home became the main site of water use, allowing people to consume frequently and on demand. In sum, these infrastructures ushered in a new era of intensified and individualized water consumption.

There are several reasons for the slow rate of change compared to the transportation or information revolutions. While railways and telegraphs were built on open spaces, water and sewage pipelines navigated dense urban terrains, making their introduction time consuming and dependent on negotiation with the public. Relatedly, the introduction of water systems required reconfiguring all elements of the built landscape, from street grids to residential architecture. While preindustrial cities were full of narrow, winding roads and dead ends, the introduction of piped systems contributed to the dominance of straight lines and wider streets in urban planning. Rather than merely an aesthetic ideal or a matter of business and political utility, the straight line was an engineering imperative that facilitated laying water, gas, and drainage pipelines. As Hugo noted, "if the geometrical line has a place anywhere it is certainly in a city's sewerage system. Here, everything must conform to the shortest route."[8]

Meanwhile, connecting residential buildings to water systems raised another set of issues. Preindustrial houses were built with materials and designed in such a way that could not support heavy water usage. Installing indoor plumbing and water fixtures in these houses posed technical and economic challenges. Often, the entire building had to be torn down and rebuilt using new materials like mass-produced pressed bricks and reinforced concrete to be connected to water and drainage. Thus, achieving universal service coverage depended on the life span of buildings and the rate at which they were demolished and rebuilt. In some cases, this process took nearly a century to complete.

But the most important reason for the transformation's slow pace was that it depended on the combined action of two infrastructures. Contrary to common sense assumptions, mechanized water supply and drainage developed at different times and along independent paths before becoming connected. The former was installed in many cities around the world during the first half of the nineteenth century, while the latter was improved and standardized toward the century's end. For instance, in London, the largest city of the nineteenth century, water supply networks were well established by the 1810s, but the construction of the sewer network began in 1858, to be completed nearly two decades later.[9] In Cairo, the water distribution network began operating in 1870, but the city waited until 1915 for mechanized drainage. As will be seen, accounting for this infrastructural interval is crucial for understanding how the revolution in water consumption unfolded.

The fact that water supply and drainage developed independently is unintentionally reflected in the scholarly literature. While studies have shed considerable light on their general impacts on public health, bodily hygiene, and modern subjectivity,[10] they have tended to focus on each system in isolation. There are excellent histories of water supply, but they largely overlook what happened to the water after it was dirtied by human consumption.[11] Similarly, there are outstanding histories of sanitation and sewerage, but they mostly ignore how water was sourced, distributed, and used before entering the drain.[12] In some exceptions, water supply and drainage are examined within the same framework, but the framework itself is often too broad, including all other urban infrastructures.[13] In others, the effect of these systems on patterns of water consumption is examined only partially, without fully articulating their novel character and radical implications.[14] This has so far hindered the full recognition of the significance of water systems to the making of global modernity. It has also obscured their comparability to infrastructures of the modern fossil fuel complex.[15] If the latter increased human dependence on coal and oil, with enormous environmental consequences, they fundamentally reshaped human dependence on water, ushering in a new environmental order. *Pipe Economics* seeks to address these issues. By studying water and drainage systems in Cairo, it clarifies the global shift prompted by their combined operation and highlights their broader impacts on modern society.

Water and the Making of Modern Cairo

Considering this global story from Cairo, it is clear that the introduction of water and drainage networks was bound up with broader transformations in the city's geography, economy, politics, and everyday life. Before the arrival of piped water, the city featured an elaborate system for sourcing, distributing, and accessing water that was dispersed across the urban landscape. Like other Egyptian cities, Cairo's supply relied on the Nile, canals, ponds, wells, and springs. Carrying and delivering water was a time-honored profession back then. There were six independent guilds that provided service to different areas, five of which specialized in bulk transportation and one in sale of water by the cup. The carriers' most distinctive implement was the waterskin—commonly made of animal hides—into which they loaded water (Figure 2). At that time, the most important places for accessing water were communal. Public baths, fountains, and mosques were not only central to the city's social, educational, and religious life, they also housed shared water and sanitary facilities that residents used on a daily basis. Located at the city's main thoroughfares, these institutions provided community service for free or in exchange of fees, with most of them being financed by wealthy patrons and charitable endowments.

The creation of the Cairo Water Company (in 1865) marked a pivotal moment in the economic history of the Middle East, as the region was assimilated to the dynamics of global capitalism before its formal annexation by European empires. Assimilation came mostly in the form of concessions, monopolies, and other corporate agreements with foreign investors. In 1854, the Egyptian government granted the French diplomat Ferdinand de Lesseps a century-long concession to build and exploit the Suez Canal.[16] Over the course of the next decade, the government

FIGURE 2. Water carriers at the Nile, Cairo, Egypt, 1896. Library of Congress.

approved similar concessions for the installation of water and gas supplies in Cairo and Alexandria, with the French engineer and businessman Jean-Antoin Cordier securing water contracts in both cities.[17] Egypt was then a semi-independent state ruled by descendants of Mehmed Ali Pasha (r. 1805–1848) while still nominally part of the Ottoman Empire. However, these economic entanglements paved the way for British occupation in 1882, which lasted until 1922.

Rather than a mere accessory to modern Cairo, the water company played a leading role in making it. Around 1868, Khedive Isma'il (r. 1863–1879) asked the company to help develop the large area between historic

Cairo and the Nile, which had remained sparsely populated until then due to its susceptibility to flooding.[18] This area became the nucleus of the modern city—now known as Khedivial or simply Downtown Cairo. In contrast to the alleyways of the city's historic neighborhoods, the modern city was distinguishable by linear streets, open squares, and the water pipes laid underground. From an infrastructure perspective, laying pipelines and building streets were complementary. After digging trenches and assembling the pipes, the company built the roads on top. Once the pipes were laid, Cairo's streets acquired a certain degree of permeance, since changing the layout would incur large costs in order to modify the infrastructure. In this sense, hydraulic engineering and urban planning were interdependent from the outset.

The changes introduced by the new network were gradual. At the beginning, the company planned to serve primarily Cairo's elites—the prospective residents of the new area under development. At the same time, it set up public standpipes in the old neighborhoods, where the traditional system based on carriers and shared facilities also continued to operate. A cholera epidemic in 1883 created the conditions for the company's role to expand. Before the mid-nineteenth century, disease was customarily understood to be linked to air quality, but with the development of the germ theory, attention gradually shifted towards water and waste as the main vehicles for pathogens.[19] Still confused about the theory, British colonial authorities implemented some public health measures in 1885, including forcing several large mosques to abolish the traditional ablution basins and to install water faucets in place. Only three years into the occupation, this brash intervention by British officials into Egyptians' daily routines and Cairo's most sacred spaces elicited the public's anger and ignited a theological debate about the suitability of piped water for religious ablution. The debate was settled quickly based on a *fatwa* (legal ruling) by the Hanafi school of Islamic jurisprudence, one among the big four in the Middle East. Since then, a faucet has been called *hanafiyya* in colloquial Egyptian, meaning *halal* (fit for ritual use) according to the Hanafi school.

In 1897–1907, Egypt witnessed an unprecedented cycle of boom and bust, a significant part of which was driven by the housing and construction industry in the country's largest cities.[20] The Cairo Water

Company benefited directly from the rapid pace of urban development. During that decade, new suburbs were built with enhanced water and sanitary amenities, and upscale neighborhoods were demolished to make way for five- and six-story buildings that required supplying more water with added pressure. As the built environment was reconfigured to better accommodate domestic water consumption, faucets, flush toilets, bathtubs, and kitchen sinks began to enter the Egyptian home. At the same time, economic boosterism was so wild that the water company embarked on a project to abandon the Nile and draw water from underground sources, promoting the speculative hypothesis that the country sat on a "hidden Nile" waiting to be tapped. The project failed amid widespread complaints against the new supply, adding to the company's record of unpopularity.

On the whole, however, the effects of the boom should not be exaggerated. By 1907, only 15 percent of the city's buildings were connected to the water company, and most of these buildings were geographically concentrated in new areas. Universalization of private water consumption pushed against two main barriers. First, the price of water service excluded the vast majority of residents. According to an observer with insider government knowledge, the company's service fees were "at least ten times higher than that of any similar establishment in Europe,"[21] where private water businesses were regulated by governments. Second, in the absence of mechanical drainage, only so much water could be used at home before causing flooding in the basement. During the boom, some landlords found themselves paying such a high price to drain wastewater that they ultimately shut off the water supply on tenants.

Cairo's drainage system was built in 1915 after more than twenty-five years of study and planning. The main reason for such a lengthy period of preparation was economic: mechanized sewerage was a highly expensive technology. In 1890, the project was slated to cost in excess of half a million Egyptian pounds—"an enormous expenditure" according to the colonial government's financial advisor.[22] A few years later, the projected cost had doubled to one million pounds. By 1907, the price tag had escalated to two million pounds, same as the Aswan Dam of 1902.[23] When the system was inaugurated, the total cost reached 2.3 million pounds, making it the most expensive engineering project ever undertaken in an Egyptian city up to that time. The government financed the system's main components, while homeowners indirectly defrayed the cost of maintenance and operations by paying higher property taxes.

Nearly one-third of the drainage project's budget was allocated to road reconstruction, which was necessary after installing the network. This aspect of the project provided an opportunity to address various issues of significance to urban planning, including traffic congestion and housing shortages, both of which were outcomes of the rapid, uncoordinated urban growth that accompanied the building boom. These efforts culminated in a comprehensive urban planning scheme designed to regulate Cairo's growth for the next fifty years. However, the plan's timing could not have been more ironic. As colonial engineers were finalizing the details, the 1919 revolution erupted in the streets, sweeping aside British rule and leading to the establishment of an independent monarchy that ruled Egypt until 1952.

The construction of the sewage system in 1915 removed the last obstacles needed for the water consumption revolution to take hold in Cairo. Over the next few decades, an increasing number of buildings were connected to water and sewers. By the late 1940s, Cairo was almost universally covered, and the effect on water usage was substantial. The original sewage system was designed to accommodate an estimated water consumption of forty-five liters per day per capita. By 1939, however, average consumption had reached ninety liters and was projected to increase further in the future.[24] In this way, the sewers acted as a powerful catalyst allowing the new regime of water access to dominate. Meanwhile, the old regime based on carriers and shared facilities was increasingly seen as an outdated mode of living. Set in this era, Yusuf al-Siba'i's novel *Al-Saqqa Mat* (The Water Carrier is Dead) is not only a story about a dying carrier but also about a dying profession and a dying world.

The Cairo Water Company emerged as one of the biggest winners in this urban transformation. With a monopoly on water supply, the company took full advantage of the increasing demand for water services. The most important tool used by the company was the water meter, which was made mandatory for all new customers in 1920. While the company argued that it was a measure to curb waste, the meter generated record profits for the company's shareholders. After a decade of simmering public resentment, the new policy was challenged both in the courts and in the Egyptian parliament. Increasingly, the water company was viewed as a glaring example of predatory foreign monopolies, of which the Suez Canal Company was the arch-villain. These companies, it was argued, exploited Egyptian resources and the public without benefiting the country's economy. As will be shown, the campaign against the water company helped define the anti-monopolistic mood on the eve of the 1952 revolution, which brought Nasser to power. His nationalization program was, in this sense, a continuation of a political trend that began decades earlier.

Learning from Engineers

To fully understand the transformative effects of water infrastructures, it is important to examine the engineers behind them. Contra common perceptions, engineering has never been a purely technical field, divorced from political, social, economic, and environmental concerns. Rather, it has always been bound to political priorities and economic interests, mediating relationships between states, markets, societies, and nature. *Pipe Economics* examines the practices and thinking processes of engineers working in Cairo, showing how they illuminate the water consumption revolution and its economic consequences for both the public and the water monopoly.

At the most basic level, engineering is the design, construction, and operation of technical systems.[25] These systems may range from ancient pyramids and irrigation canals to modern AI-powered robots and space telescopes probing the deep history of the universe, and from physically imposing infrastructures like dams or national power grids to intangible or microscopic technologies such as computer software and gene editing tools. Academic studies have already shed much light on the profession and its practitioners.[26] One strand of the literature focuses on the professionalization of engineers, often exploring their education, training, career paths, and the development of professional associations in national contexts.[27] Another strand has looked at engineering as the key for understanding the relationship between technological and social change.[28] A third strand has explored engineering as a distinct mode of knowledge that is practical, hands-on, and derived through a process of trial, error, and feedback.[29] *Pipe Economics* draws on these studies to explore the engineers who designed, managed, and assessed water systems in Cairo.

During much of the period here under study (1865–1957), the engineering profession was structurally divided by ethnicity and shaped by the colonial context. The engineers who designed and built Cairo's water infrastructures were mostly foreigners, predominantly from Britain, France, and Germany. Reliance on foreign engineers began during the Khedivial period and in the context of building the Suez Canal and urban development projects. Cordier, the founder of the water companies in Alexandria and Cairo, is a case in point. However, the role of foreign engineers expanded greatly under British rule, with the Public Works Ministry dominated by British engineers and advisors. During that time, Egyptian engineers were systematically sidelined from large projects and relegated to minor roles.

This marginalization bred forms of resistance, dissent, and political violence. In 1893, Egyptian engineers launched *al-Muhandis* (*The Engineer*), the first Arabic-language engineering journal in Egypt, in response to William Willcocks, a British engineer who argued that classical Arabic was unsuitable for knowledge production and dissemination. Around 1905, *al-Muhandiskhana* (the Egyptian Engineering School) became a hotbed of anticolonial organization, with several of its graduates arrested in 1910 in connection with the assassination of Prime Minister Butrus Ghali, a political crime that sent shockwaves through the country. Because they had no say in the halls of government, Egyptian engineers often voiced their critical opinions in the nationalist press, with one whistleblower warning against certain features of the sewage system project when it was under construction. Following the 1919 revolution, the balance of power shifted toward Egyptian engineers, who finally assumed leadership of the Public Works Ministry.

Pipe Economics resists the tendency to cast engineering as a one-dimensional tool of empire, which is often the result of well-intentioned efforts to decolonize the profession. It is one thing to recognize the overarching colonial context that has shaped the practices and ideologies of foreign engineers working in Egypt, and quite another to reduce everything they did and wrote to sheer instrumentality. Naturally, some British engineers did the bidding of the colonial government, designing uneven and inadequate schemes that would have inconvenienced most Egyptians. Others, however, offered insights that merit careful consideration for what they reveal about the city's geography and economy. What follows is an attempt to show that a critical approach to engineering can take colonialism seriously while still learning from engineers.

One of the mainstays of the history of science and technology is to defamiliarize the familiar by challenging linear, progressive narratives of scientific discovery and technical innovation.[30] Historians of technology have shown that the development of technical innovations was often a messy process involving competing designs, adjustments, contingencies, and social and political contestation. Once a technological design became dominant, however, it often erased the history of its own becoming, giving the illusion that it was born fully formed and armored like the Greek goddess Athena. By foregrounding the engineers who designed Cairo's water systems, this book seeks to take the reader back to an era when water and sewer networks were unfamiliar technologies still being configured and standardized. Because of their novelty, their effects were subject to wide-ranging debates and educated guesswork not only among laypeople, but also among the experts themselves.

As engineers studied Cairo's sewage system project, they endlessly debated how much water Egyptians used under the current system of carriers and shared facilities, and how much they would want to use in the future, after drainage was installed. The issue was not merely theoretical or speculative. In fact, it was of critical economic importance, because it directly influenced the overall cost of the scheme—the colonial government was initially unwilling to make a large investment solely for the benefit of Cairo's residents. In attempting to answer these questions, engineers scrutinized every aspect of the city's hydrology, examining the water company, carriers, cisterns, mosques, fountains, baths, and domestic consumption. Following these debates reveals how engineers theorized the water consumption revolution while designing the infrastructures that would make it possible. In this way, they were the first to observe the transition toward the new hydrological regime based on private, on-demand water access, anticipating its eventual emergence as a global standard.

Pipe Economics draws on engineering to explore the economics of water management in Cairo. Engineers do not only design and build technical systems; they also analyze and theorize their economic dimensions. In the mid-nineteenth century, governments and private companies across the globe employed engineers to manage the financial affairs of canals, bridges, railways, and water systems. With the rapid expansion of these costly infrastructures, economic analysis became a routine task that engineers were trained and expected to perform, often under the rubric of cost-benefit analysis. In some notable cases, engineers authored treatises that formalized their practices and addressed key issues. One of the most influential works in this respect is *Railway Economy* by Dionysius Lardner, a London-based Irish engineer and popularizer of science and technology.[31] Published in 1850, Lardner's book provided a comprehensive survey of various economic issues related to the railway industry, including the role of railroads in economic development, trends in passenger and freight traffic, costs of construction and maintenance, and pricing strategies for railway companies and governments. Since then, the field of railway economics has generated interest not only among engineers but also among academic economists.[32] In the twentieth century, engineering economic analysis further expanded to encompass the totality of industrial systems and their connected supply chains. In *The Engineers and the Price System* (1921), the American economist and sociologist Thorstein Veblen borrowed from contemporary engineers to critique the predatory nature of business control over economic life.[33] Building on this tradition at the crossroads of the two fields, *Pipe Economics* highlights the economic thinking of Cairo's hydraulic engineers and draws on their insights to better understand the political economy of water management in the Egyptian capital.

Notably, engineering analysis underscores the common economic features shared by all infrastructures. To begin with, infrastructures do not cater to a limited number of users; rather, they supply goods and services to whole new markets of mass consumers. Civil engineers were often the first to conduct market research, assessing the potential economic benefits of these infrastructures before governments and private companies committed to large investments. Relatedly, infrastructures are characterized by the preponderance of "fixed capital"—substantial investments in durable, immobile assets—which makes them inherently risky. If a project fails to deliver the expected benefits, investors and governments would find it challenging to repurpose assets for alternative uses.[34] While some assets could be easily shipped and sold overseas, rails, canals, pipelines, and bridges "cannot be exported body and soul," as Karl Marx noted.[35] Finally, infrastructures often operate as "natural monopolies," meaning economic sectors where business competition is considered impractical or undesirable. In these cases, a single network typically serves a large number of users, and the high cost of building a competing network generally deters potential entrants.[36] In exploring the economics of Cairo's water systems, this book seeks to contribute more broadly to the political economy of infrastructures.

Structure of the Book

The events presented in this book span nearly a century and four political regimes. Rather than providing an evenly distributed account of this history, the chapters focus on pivotal events that altered, or sought to alter, the flows of clean and waste water in the Egyptian capital, exploring the role of government, business, and disease in driving urban change, and examining how people responded to these developments. While the story is cumulative, each chapter presents a separate case study in infrastructure, with different themes situated at the intersection of urban, environmental, economic, and science and technology studies. The book begins and ends with chapters on the Cairo Water Company, while the middle chapters detour into the city's sewage system, examining several aborted projects, and explore the role of the construction and housing industry in Cairo's hydrological transformation.

Chapter 1 surveys the city's old hydrological order, examines the formation of the water company and its role in urban planning, and investigates the cholera epidemic of 1883 and its consequences for water supply and sanitation. Chapter 2 examines a sewage system project from 1890, focusing on a debate among colonial engineers about the project's economic and technical viability. The debate, I argue, shows that once a drainage system was built, demand for domestic water service would soon increase. But because the project was not implemented, this prediction remained a theoretical possibility waiting to be tested on the ground. In Chapter 3, I examine another failed drainage scheme, but this time I explore its relationship to the British colonial regime's efforts to modernize the country and upgrade the bureaucratic apparatus. To finance the project's operation and maintenance, the government planned to raise the house tax from 8.3 to 10 percent. The attempt to raise house taxes in Cairo foreshadowed an initiative by the colonial administration to update the whole country's taxation system using new cadastral maps.

Chapter 4 turns to the boom of 1897–1907 to examine its effects on both the construction sector and the water industry. It shows how real estate developers marketed new housing as "modern" and "healthy," partly on account of improved water and sanitary amenities. It also examines the water company's ill-fated attempt to switch the source of water supply from the Nile to deep wells. In Chapter 5, I examine the design and implementation of Cairo's drainage project, focusing on its implications for urban planning as well as municipal and national politics. As the government prepared to raise property taxes, critics of colonial rule protested against taxation without political representation, demanding the creation of an elected Cairo municipality. Chapter 6 returns to the Cairo Water Company to examine the mounting legal and political challenges it faced from the 1930s onward. The company had mandated the use of meters in 1920 as a shrewd business strategy, correctly predicting the rise in demand for domestic water service following the construction of the sewage system. In 1937, the government hired a British water engineer with expertise in water pricing to arbitrate between the company and the public. Focusing on this arbitration, I show how the company's predatory behavior was exposed and challenged. The book's conclusion traces the arc from nationalization in 1957 to the resurgence of water privatization since the early 1990s.

ONE

Privatizing Water

The Water Company and the Transformation of the Urban Landscape

The Cairo Water Company emerged from a powerful alliance between private business and political power, with Nubar Pasha at its center. A statesman of Armenian descent, Nubar played a leading role in Egyptian politics, serving as premier for three terms (1878–79; 1884–88; 1894–95). Educated in France, he earned a reputation as a seasoned diplomat, legal reformer, and one of the wealthiest men in Egypt.[1] He is particularly noted for securing important loans for the country, aiding Khedive Isma'il Pasha (ruler of Egypt from 1863 to 1879) in negotiating greater autonomy from the Ottoman government, and establishing the Mixed Courts.[2] However, Nubar is also notorious for being a businessman with a knack for self-enrichment.[3] In 1865, he was appointed Minister of Public Works and granted nearly unlimited powers.[4] As he reflected in his memoirs, he found the role particularly appealing, viewing Egypt as "a large factory for public works."[5]

In that same year, the Egyptian government granted French engineer Jean-Antoin Cordier the exclusive right to distribute water in Cairo for ninety-nine years—his second water concession after Alexandria.[6] Soon afterwards, Cordier founded the Cairo Water Company. The concession contract—*Firman Authorizing the Establishment of a Nile Water Distribution System in Cairo and Its Suburbs*—was signed by Nubar Pasha in his capacity as the Minister of Public Works and the government's representative. As it turned out, however, Nubar was also Cordier's most important business partner—with 12,000 of the water company's 68,000 original shares—and the first chairman of the board.[7] This conflict of interest was unknown to the public at the time, though it was an open secret in business and government circles and remained so for decades.

The company's founding coincided with a series of ambitious reforms and development projects by Isma'il Pasha. The Khedive invested heavily in the construction of the Suez Canal, the renovation and expansion of Cairo and Alexandria, the establishment of a sugar industry, and the extension of irrigation and transportation infrastructure throughout the country. However, much of this progress was funded by borrowed money. European financiers were eager to provide capital, technology, and expertise in exchange for exclusive, long-term privileges. As the Khedive's ambitions continued to outpace the state's resources, Egypt's ability to service its growing debt diminished, ultimately leading to bankruptcy in 1876.[8]

This chapter provides an overview of Cairo's traditional system of water distribution and access, chronicles the early history of the Cairo Water Company, and examines its multiple impacts on the Egyptian capital. As will be shown, the old system was based on common water resources, water-carrier guilds, and shared facilities that were available in every neighborhood. Following the water concession, Cairo's water supply became subject to a private property regime in which a single company wielded huge influence. At the same time, the water carriers found themselves squeezed between the company's monopoly power and increasing government regulations. Viewed from a broad historical perspective, this was the moment of "primitive accumulation" in the classic Marxist sense, where common resources were appropriated to establish the rules of private property, forcibly separating communities from their means of subsistence and production.[9] During the following decades, the company further secured its position through strategic agreements with the government.

Before COVID-19, there was cholera. The disease wreaked havoc on the entire globe throughout the nineteenth century, striking in six pandemic waves. The 1883 cholera epidemic in Egypt was therefore neither an isolated nor an exceptional event. However, it took place at a particular moment in the development of modern medicine when water supply and sanitation came under intense scientific scrutiny.[10] Previously understood as miasma, or "bad air," cholera was beginning to be linked to a specific contagion suspected of spreading via contaminated water—in fact, the proof emerged partially from Egypt. Much like COVID, cholera was intensely politicized and surrounded by rumors concerning its origin and mode of transmission. Even as British public health experts rejected the germ theory of disease, they nevertheless found a convenient rationale for intervening in Cairo's largest mosques, ordering the replacement of traditional ablution basins with water taps. Without the 1883 epidemic, it is not clear when water taps would have been installed inside traditional facilities. But the evidence suggests that it was only a matter of time. Drawing on the work of historian Khaled Fahmy, the following account underscores the centrality of public health to understanding urban change, showing how cholera left a lasting imprint on urban geography and proved consequential to the consolidation of the city's water company.[11]

Before the Company

Using water was rarely a private activity in historical Cairo. People drank, cooked, bathed, and performed ritual ablution together in the same places where they socialized, studied, and prayed. Born out of centuries of gradual development, the old system of water access and consumption was well integrated into the city's social and political life. It was also particularly adapted to the traditional system of sanitation, in which waste recycling—into fuel, building materials, fertilizers, and gunpowder—was the common order of things. Displaced by industrial and colonial imperatives, the old systems of water and waste management thus offer vivid examples of preindustrial forms of sustainability and circular urban metabolism.

The Nile was the primary source of fresh water for all of Egypt. Each summer, rainfall in the Ethiopian highlands caused a significant increase in the Nile's flow. From June to mid-September, rainwater rushed down the rugged mountains to form the Blue Nile, one of the river's two main tributaries. By late summer, the first signs of flooding reached Egypt. Over the next three months, the water would sweep through the entire valley, overflow the riverbanks, and inundate the adjacent floodplain. In January and February, the river receded and regained its calm but continued to flow full. From February to May, the water levels gradually decreased, reaching their lowest and most stagnant point by mid-June. Two months later, the cycle began anew with another surge of fresh water from the south. This annual cycle sustained life in the Nile Valley for thousands of years.

Cairo experienced the same flood cycle that swept through the entire valley. This is why the city's old quarters were built away from the river and on the elevated land at the desert's edge. During exceptionally strong flood seasons, water covered extensive areas, sometimes reaching the city gates and spreading as far as the Pyramids of Giza on the left bank (Figure 3). The Mamluk historian al-Maqrizi described a high flood that took place in 1323 (723 in the Islamic calendar), in which a large area between Bab al-Bahr, one of Cairo's western gates, and Bulaq, then a separate port town, was submerged. The water was so high that people could ride boats and pick fruits from submerged orchards.[12]

The city used to be crisscrossed by canals and ponds. One of the most significant canals was the Khalij, a narrow waterway that cut through the city, starting from the south at Fumm al-Khalij (literally the mouth of the canal) and extending northeast beyond the city limits (Figure 4). The Khalij was dammed for most of the year, running low or even dry between January and August. Opening the canal took place in mid-August and was marked by a popular ceremony accompanied by lively street celebrations. Another important canal, the Bulaqiyya, ran between Bulaq and Cairo. Smaller canals fed the city's large ponds, most notably al-Azbakiyya, al-Fil, and al-Saqqayin (the Water Carriers' Pond).

Water quality varied with the Nile's seasonal cycle. During the peak of the flood (late August to the end of September), the river turned red due to the abundance of silt. This "red water" was typically left to settle in clay jugs before being used for drinking and cooking. Conversely, the water was considered purest during the Coptic month of Tuba

FIGURE 3. Nile flooding near the Pyramids of Giza by Émile Bayard (1890–95). Bibliothèque Nationale de France.

(January 9–February 7), when the river flowed smoothly and steadily. By the second half of June, at the height of the ebb season, the Nile turned green due to the presence of a large mass of microscopic algae. This "green water" was deemed foul or even poisonous if consumed untreated. Therefore, beginning from the 10th of Bu'una (June 16), it was common practice to boil the water with medicinal substances like alum and apricot or peach kernels, allowing it to settle in jugs to achieve acceptable taste and purity.[13]

Carrying water was a vital profession in preindustrial Cairo. Like other craft and service workers in Egypt, the carriers were organized into guilds overseen by a master, who acted as a mediator between workers and the government, collected taxes from members, set prices for goods and services, and regulated entry into the profession.[14] By the mid-nineteenth century, Cairo had six guilds dedicated to water distribution—one specializing in selling water by the cup and the others in bulk transportation. During the flood season, carriers commonly

FIGURE 4. Khalij Canal, c. 1895. Bibliothèque Nationale de France.

drew water from canals and lakes. Throughout the rest of the year, they fetched water directly from the Nile at designated locations. Carriers loaded the water in skins made of animal hides, and used draft animals for long-distance transportation. Delivery operated on a subscription system: carriers either marked doors with lines indicating the number of waterskins delivered or provided customers with a number of blue stones, subtracting one upon each delivery.[15] Despite the widespread use of water wheels in Egyptian cities, manual labor remained essential, as water still had to be carried and delivered to points of consumption. One of the city's oldest water features is the twelfth-century aqueduct built by Sultan Salah al-Din al-Ayyubi, which raised water from the Nile to his Citadel, the traditional seat of government.

Since communal water usage was the norm, most of the water transported was delivered to public baths, fountains, and mosques. Public water facilities were typically well distributed along main roads and major intersections. They could be found either as standalone structures or clustered together near schools, markets, hospitals, and caravanserais.[16] Financially supported by charitable endowments, most of these facilities were free to the public—or at least to a large segment of the Muslim male population—while others charged modest fees for their services.[17]

The public bath was a multifaceted institution of vital significance to social life, public hygiene, food supply, and waste recycling. Cairenes of all classes visited baths not just out of necessity, but also to relax and socialize. Baths were lively spaces for sharing news and for arranging marriages and business deals.[18] Notables celebrated holidays with meals, music, and dance in these establishments. Some baths were gender-specific, while others operated on alternating shifts. Foreign observers admired these baths, comparing them favorably to European public services.[19] By the beginning of the nineteenth century, Cairo had about one hundred baths, with fifty-nine still operating a century later.[20]

Public baths shared common design features. Built with thick, windowless walls to retain heat, they were always surrounded by other buildings. The entrance was small and ornamented. Inside, the layout included three main rooms, progressing from cold to hot. The first room, the reception area, was used for undressing, resting, and socializing; it was usually spacious and decorated with motifs and calligraphy. The second room served as a transition before entering the hot spa room. The final room often featured a central bench and a dome with colored glass. The hot room was surrounded by side rooms and plunge pools. Bathers immersed themselves in hot water and received massages before returning to the reception hall to collect their belongings and pay the bath manager.[21] Fees varied according to the facility's opulence and neighborhood wealth. Baths typically had their own wells but also sourced extra water from carriers, especially in the summer.

The water heating system of public baths offers a glimpse into recycling and culinary practices in historical Cairo. The furnace, typically located above the plunge pools, was hidden from view and accessible only from a separate street. Covered in salt to conserve heat, the furnace featured small pipes that directed hot water to the pools. Public baths commonly burned trash for fuel. In 1909, for instance, they bought nearly a third of the city's trash from street sweepers, who also formed their own professional guild.[22] Inside the furnace, large pots of fava beans were slow-cooked before being distributed to food markets. The ash from furnaces was used as a key construction material for plastering walls.[23]

Another notable public facility was the water fountain, which provided free drinking water for people and animals. Established by rulers and wealthy patrons, these fountains were symbols of piety and dedication to public service. Cairo's fountains were often ornamented with motifs and inscriptions bearing the founder's name, verses from the Quran, or expressions of piety.[24] Wealthy women also contributed to this tradition; for example, the fountain of al-Sitt Saliha (Lady Saliha), built near the Khalij Canal in 1741 (1154 in the Islamic calendar), carried her name and the inscription: "I built a cistern for God in hope of blessed returns and good fortune."[25] Due to their large storage capacity, fountains were crucial during the ebb season. At the beginning of the nineteenth century, there were about three hundred public fountains in Cairo, with around two hundred still operating in the 1880s.[26]

Public fountains were typically composed of three separate levels. The first was the underground cistern, which was made of marble or moisture-resistant stone and filled at least annually by water carriers or from nearby canals. The second was the distribution chamber, built at street level or slightly above. The fountain overseer would close this space at night after washing and perfuming drinking jars. Ottoman-era fountains featured a tap-like fixture for easy access to water jugs.[27] Lastly, a small school was usually located above the distribution room, where poor children were taught reading, writing, and the Quran.

The mosques of Cairo—known as the city of one thousand minarets—need no introduction, but their role as public sanitary facilities might be less familiar to contemporary readers. Beyond their function as centers for worship, education, governance, and socializing, mosques were crucial for bodily hygiene, as they comprised facilities for ritual ablution. They also served as public latrines and were used daily by most men, with some mosques offering facilities for women as well. Similarly, churches and synagogues provided sanitary services for Christians and Jews in their respective neighborhoods. By the late nineteenth century, Cairo had nearly three hundred mosques.[28]

Courts of ablution, often centrally located in mosques, featured marble or stone basins. While mosques typically had their own wells and cisterns, they frequently relied on water carriers. The basins were regularly replenished with fresh water based on usage. Mosque latrines were located in a separate room and contained traditional squat toilets. Before piped water, small clay jugs were used for cleaning and flushing. Toilets were connected to drop pipes leading to a percolating cesspool or a nearby body of water. By the late nineteenth century, al-Azhar Mosque's latrines served over eight thousand visitors daily.[29]

Percolating cesspools were well suited to the city's geography and preindustrial needs. These underground receptacles were often unlined and built with porous materials, which allowed liquids to seep into the soil while retaining solids. This method reduced the need for frequent emptying by leveraging the soil's absorptive qualities and the city's dry climate. As a result, mosque cesspools were typically emptied every two to four years. Emptying cesspools was the domain of waste carriers, who also formed their own professional guild. Carriers first added sand to the cesspools, then used shovels and buckets to transport the waste to dumpsters outside the city. Waste was typically left to dry before it was harvested and used as agricultural manure. Al-Maqrizi described the frequent bargaining between waste carriers and customers. Since carriers were usually called when cesspools were full, this information was often used against their customers "to exaggerate the fee and multiply its value."[30]

Domestic water usage was the exception rather than the rule. Nineteenth-century houses were simply not designed for large volumes of water consumption. Built with thick limestone, the ground floor was used for commercial purposes on busy streets or residential use in quieter areas. The second and third floors were made of lighter bricks and used for living.[31] Most houses had squat toilets connected to bottomless cesspools. Rich households owned cisterns, though rarely large enough for private baths. Wells were common and were typically five to six meters deep—enough to reach underground water during most of the year. But the water obtained was primarily used for cleaning and gardening, less for cooking, and even less for drinking.[32] Wealthy homes featured an elaborate ventilation system with internal courtyards, wind catchers, and narrow corridors to renew air.[33]

The Water Concession

The old regime of water access and consumption began to come to an end in 1865, when the Egyptian government granted the Cairo water concession. As a result of this concession, the water of the Nile was no longer a common resource to be managed by local communities, but rather a new commodity to be exploited for profit on an industrial scale. Once carried on the backs of men, women, and animals, water was now to be distributed by means of machines, pumps, pipes, and taps. Previously counted in waterskins, water was now to be measured in liters and cubic meters.

Contrary to conventional wisdom, the privatization of water resources is not a recent development tied to late twentieth-century neoliberalism.[34] In fact, it has deeper roots that stretch back to the early nineteenth century when private companies built the first water-supply networks in major European cities. London, for instance, was home to eight private water companies that sometimes competed aggressively and others agreed to divide the city into exclusive zones of control.[35] The rapacious behavior of these water companies led an economist like John Stuart Mill, known as a promoter of laissez-faire economics, to consider water supply an unsuitable domain for business competition, arguing that it was best managed as a public monopoly and overseen by municipalities and government-appointed boards.[36] In his view, water distribution was a case of "natural monopoly," where competition does not yield the assumed positive outcomes but instead harms both the public and private enterprise. This concept was later expanded by other theorists to include sectors like gas, railways, drainage, and electricity—industries typically characterized by a single network or line serving a large number of users.[37] In such cases, competition led to irrational results, forcing new entrants to invest in building costly infrastructure where one already exists. Mill's views shaped government regulation between 1847 and 1902, when the British Parliament enacted a series of laws to regulate the quality of London's water supply, limit company profits, and ultimately place water companies under the control of a government board.[38] Contrast this with the situation in Egypt, where the Cairo Water Company emerged from its inception as an uncontested private monopoly, facing no significant political challenges until the 1930s.

The Cairo Water Company was backed by a small group of prominent businessmen. Jean-Antoine Cordier, the original concessionaire, was born in 1810 in Béziers, France, and trained as a hydraulic engineer. His father also specialized in water installations, but mostly in France rather than abroad. Other than Nubar Pasha, the founders included Jules and Jean-Baptiste Pastré, Marseille businessmen and investors who had lived in Alexandria for decades. The Pastré brothers were instrumental in founding the Anglo-Egyptian Bank in 1862 and were active in trading textiles, cereals, and construction equipment. Another founding member, Demetrius Nicolopoulou, was a cotton trader of Greek origin based in Marseille, who also held investments in the Bank of Alexandria (established in 1857).[39]

In theory, the Cairo Water Company held a monopoly not over water distribution in general but specifically on piped water supply. Article 2 of the concession contract clarified this point, stipulating that it did not infringe upon "all other means of supplying water"—a dismissive reference to the existing carrier-based system.[40] In practice, however, it was the beginning of the end for manual water transportation. Since no one expected the whole city to be connected to the company mains immediately, carriers would still be necessary to distribute water, especially in areas not fully covered by the network.

The price of water was subject to rights and obligations stated in the concession contract. The company was authorized to sell piped water at a maximum rate of "ten paras per kirbé [*qirba*, waterskin], estimated at one-fifteenth of a cubic meter, or approximately sixty-six liters."[41] Several points warrant attention here. First, although water carriers were not mentioned by name, their implement—the waterskin—was used as a crucial reference point for converting between traditional and modern systems of measurement, between average-sized carrying loads and specific quantities expressed in liters and cubic meters. Second, the para was then used in various provinces of the Ottoman Empire. After it was discontinued in 1885, the water rate was simplified to one French franc (thirty-eight millimes) per cubic meter.[42] Third, while the contract claimed that the price of water was "half of that currently paid,"[43] the comparison is misleading or disingenuous, as it suggested that the company was cheaper than the carriers. In fact, the rate charged by carriers pertained only to home delivery. Standardizing this rate overlooked the fact that Cairenes typically accessed more water for little or no cost at mosques, baths, and fountains, and by drawing it themselves. Lastly, the concession permitted the company to "sell at a lower price and offer subscriptions," a clause that led to considerable controversy decades later, as discussed in Chapter 6.[44]

The company was granted additional subsidies. Expensive machinery and cast-iron pipes were exempt from local duties.[45] The government also freely granted the company a plot of land south of Cairo on which to build the waterworks. This land was formerly used for gunpowder production, before the industry was relocated to the Island of Roda. However, it retained the negative reputation associated with gunpowder making, then considered a dangerous and noxious process involving refining animal manure into saltpeter, a combustible material.[46] Relatedly, the site was in close proximity to dumpsters, animal markets, and other polluting industries. This unsanitary reputation likely contributed to the water company's decision to abandon the site before building anything, especially as a cholera epidemic hit Egypt in 1865.

The outbreak triggered widespread panic: Khedive Isma'il reportedly fled to Istanbul, while concerned Europeans evacuated the country. According to historian LaVerne Kuhnke, the epidemic brought business to a virtual standstill in both Alexandria and Cairo.[47] Thus, the company's plans were delayed for a few years. By the early 1870s, however, the company had built the network's main components. Instead of the old site of gunpowder manufacturing, the pumping station was placed in Bulaq, near the water intake point where the Nile met the Isma'iliyya Canal. The company built two reservoirs on the city's outskirts, strategically located on elevated land to provide sufficient pressure within the network. The water filtration system consisted of large sand beds placed in the desert of 'Abbasiyya (Figure 5). Upon completion, the company's network covered a substantial portion of the city.

Building Modern Cairo

To fully understand the company's significance for Cairo's history, we must go back to 1867. That was when Khedive Isma'il embarked on an ambitious project to develop the large area between the Nile and historic Cairo and to "modernize" parts of the older neighborhoods. Historians

FIGURE 5. Sand bed filters in 'Abbasiyya by Émile Béchard, 1874. Bibliothèque Nationale de France.

of Egypt have debated the broader significance of Isma'il's urban development project. According to the conventional narrative, it marks the beginning of modern Cairo.[48] One of the cornerstones of this narrative is the grid plan of the new area. With its open squares and broad linear boulevards, it has been viewed as an attempt to imitate Paris. In fact, however, it resembled not only Paris after Haussmann's renovations but so many other cities undergoing large-scale development at the time. The dominance of the straight line in urban planning and engineering reflected not only aesthetic sensibilities and political commitments, but also the need to improve air circulation and the necessity of laying infrastructures, including for water and sanitation. This connection is most evident in Cairo, where the water company installed the pipelines and built the streets of the new area in a single operation.

As is well known, part of the motivation behind the Cairo development project was to prepare the Egyptian capital for the grand festivities celebrating the opening of the Suez Canal, planned to take place in 1869. At first, the Khedive turned to 'Ali Mubarak, one of the most prominent Egyptian engineers of his generation.[49] After graduating from *al-Muhandiskhana* (the School of Engineering) , Mubarak was sent on an educational mission to France, where he studied military and civil engineering between 1844 and 1849. Later in his career, he authored several books, including the monumental *Al-Khitat al-Tawfiqiyya al-Jadida li Misr al-Qahira* (*Tawfiq's New Plans for Cairo*)—a twenty-one volume exploration of Cairo's history and topography. In 1867, Isma'il asked Mubarak to oversee several interconnected projects: 1) renovating the Azbakiyya neighborhood; 2) developing the area between Cairo and the Nile, later named Isma'iliyya; 3) drafting a master plan for the entire city; and 4) constructing wide streets in old urban areas.

Initially, the Azbakiyya neighborhood witnessed most change. By 1868, the neighborhood's pond—one of the largest in Cairo—had been drained, and plans were underway for the construction of a garden, an opera house, and a theater in place. On the other hand, the project to carve wide boulevards in old areas was downsized due to the large scale of expropriation and demolition required. Instead, Mubarak focused his efforts on three main roads: Muhammad 'Ali, Clot Bey, and al-Sikka al-Jadida (the New Road). The construction of these roads had begun decades earlier, but it was a slow process requiring the destruction of hundreds of homes as well as several mosques, monuments, and cemeteries. Mubarak did not take this lightly; rather, he saw it as necessary for improving air circulation, which he believed justified the demolitions on hygienic grounds.[50] Like most medical authorities at the time, he believed in the miasmatic theory of disease, and this influenced his vision for what needed to be done in Cairo.

Progress was slow in the large area between the historic city and the Nile, designated for the development of Isma'iliyya. This area was marked by a peculiar geography shaped by annual flooding and the presence of animals and noxious industries. Small villages existed alongside swamps, animal markets, the slaughterhouse, the tanneries, gunpowder manufacturing, dung cake makers, and mounds of rubbish that were placed to control floodwaters.[51] This landscape made it difficult to rapidly develop the area and deeply offended Mubarak's hygienic sensibilities: "As for the area of the tanneries and Bab al-Luq, do not ask about what sort of rotting materials and awful smells it contains."[52] Under Mubarak's direction, the slaughterhouse and tanneries were relocated to the south. However, Khedive Isma'il grew dissatisfied and worried that Cairo was would not be ready for the Suez Canal opening ceremony. At this point, he turned to the Cairo Water Company for assistance.

As historian Jean-Luc Arnaud has shown, the Khedive's decision to turn to the water company was driven primarily by practical considerations.[53] The company was already in the process of surveying Cairo to plan the distribution network. As a result, it possessed detailed information and maps that the government lacked. After a few months, Cordier presented a plan for Isma'iliyya, showing roads, land subdivisions, and proposed locations for government ministries and offices.[54] Over the next few years, the company first leveled the land using rubble from nearby dumpsters, then proceeded to install the water network before building the streets on top. At the same time, the government briefly considered building a drainage system, but the estimated cost was deemed too prohibitive. Instead, the government built five sewer lines to serve the main streets of the new Cairo. Notably, however, they were designed not to collect domestic wastewater but only water from street level, including washings and stormwater. These sewers discharged directly into the Nile and the canals.[55]

With much of the work done by 1871, Cordier left the water company and was succeeded by Jean-Baptiste Pierre—a French businessman with experience in waterworks in the Suez Canal zone—who served as director until 1889. Meanwhile, the government employed French engineer Pierre Grand to oversee any remaining work. In his capacity as director of the newly created Roads Department, Grand negotiated with the existing villages in Isma'iliyya, overseeing expropriation and demolition in the Shaykh Rihan Hill area as well as of the village of al-'Abit, located in what is now Qasr al-Dubara, near Tahrir Square.[56]

The first buildings to be connected to water mains were in the newly developed areas. These included government buildings, hotels, banks, and the homes of a small group of elites. The company's initial customers were members of the royal family, government officials, consuls, bankers, contractors, businessmen, merchants, and doctors, many of whom were foreigners or protégés of foreign governments.[57] This affluent clientele was the first to receive piped water at home. At the same time, the company installed standpipes in old neighborhoods to sell water to the public (Figure 6). These standpipes marked the beginning of a new role for Cairo's water carriers. Some became middlemen and retailers, selling company water on the streets or at the standpipes. Others resisted this new reality, continuing to draw water on their own, even as the government tightened regulations and eventually banned them from distributing unfiltered water. The carriers also suffered from the government's policy of filling in ponds and canals. Stripped of their autonomy and their livelihood, they resented the company and shaped the popular reception of the new water supply.

Khedive Isma'il continued to borrow in order to finance his grand schemes until the Egyptian state officially declared bankruptcy in 1876. Britain and France exploited the situation to extract political concessions from the Khedive, leading to the imposition of foreign control over Egypt's finances. In an attempt to regain solvency, Isma'il increased taxes not only on peasants but also on his allies among local notables and large landowners. These developments set the stage for the 'Urabi revolt, which Britain used as a pretext to occupy Egypt in 1882.[58]

FIGURE 6. A standpipe on Bab al-Wazir Street. Sueddeutsche Zeitung Photo / Alamy.

Cholera

Less than a year into the British occupation, Egypt was struck by a devastating cholera epidemic that caught the new colonial regime off guard. The outbreak exposed the regime's incomplete control over the administration. Although the response was swift, government experts often worked at cross-purposes, reaching contradictory conclusions. The state of disarray was partly due to lack of scientific consensus on cholera's origin and mode of transmission, but it also stemmed from the intense politicization of the issue, especially in light of its potential impact on British shipping interests. The uncertainty and fragmented response created fertile ground for rumors and conspiracy theories to flourish.

The disease first appeared in the Mediterranean port city of Damietta in June 1883, after approximately 15,000 visitors had gathered to celebrate the popular feast of Shaykh Abu al-Ma'ati.[59] The government swiftly imposed a quarantine to prevent people from leaving the city. However, cholera soon spread to Mansura, Port Said, and Alexandria, before reaching Cairo. The highest number of fatalities in the capital occurred on July 22 and 23, and by August 25, cholera had begun to recede. Official statistics reported 6,488 deaths in Cairo, then a city of nearly 368,000, though this figure was later revised to 8,000 to account for the many unrecorded deaths during the peak of the outbreak.[60] Across all of Egypt, between 80,000 and 100,000 people died from cholera.

There were four different administrations managing the epidemic. The first was the Alexandria-based International Board of Maritime Quarantine, reorganized according to the recommendations of the 1866 international sanitary conference in Istanbul.[61] This board, composed of representatives from fourteen European countries, aimed explicitly to stop cholera in Egypt or the Hijaz before it reached Europe.[62] The second administration, the Cairo-based Board of Health, dealt exclusively with internal public health administration.[63] A third government bureau, a special sanitary commission composed of local notables and consular officials, was chaired by the Prefect of Cairo Police. Finally, medical officers of the British army made critical decisions during the outbreak.

The city's air must have smelled of burning tar and sulfur, as medical authorities still believed that these fumes could purge miasma. The government disinfected markets, slaughterhouses, canals, mosques, and public baths. Mosque latrines deemed insufficiently hygienic were shut down, and rubbish was disinfected and incinerated. The homes of cholera patients were washed with lime, and their clothes were burned. Widespread sanitary inspections were conducted in the slums of Bulaq and Old Cairo, areas hardest hit by the epidemic. In Bulaq, many inhabitants were forcibly evacuated, and their houses were "destroyed by fire."[64] At the British army's behest, one group of refugees was relocated to a camp in Tura in the south, while a second, larger group was sent to a camp at the Nile Barrage in the north, bringing the total number of evacuees to 6,000. Stigmatized as sites of poverty, disease, decay, and moral depravity, slum demolition was a common practice in cities worldwide.[65]

Known colloquially as *al-shawta* (the kick) and in more refined Arabic as *al-hawa' al-asfar* (yellow wind) and *al-hayda* (a general term for disease), cholera was associated with air and changes in meteorological conditions. During the epidemic, residents of Bulaq were desperate to escape both the disease and the colonial government. French illustrator Georges Montbard, who witnessed the outbreak, captured scenes of terrified families carrying their children and belongings, huddled together on boats (Figure 7). Another illustrator, whose work appeared in *Life* magazine, vividly depicted Bulaq refugees leaving the quarter (Figure 8).

In 1883, there was no global scientific consensus on cholera's origin and mode of transmission. According to the still dominant miasmatic theory, cholera, like other diseases, could spontaneously rise up from decaying organic matter under the right circumstances. But even within this framework, some doctors believed that quarantines and sanitary

FIGURE 7. Cholera in Bulaq, 1883, by Georges Montbard. Wellcome Library, London.

FIGURE 8. Cholera in Bulaq, 1883. *Life* Magazine.

cordons were the best measures to stem infection, while others argued that these measures were outdated and unnecessarily harsh.[66] By 1860s–70s, scientists like Louis Pasteur and Robert Koch began to disprove "spontaneous generation," arguing that diseases were caused by specific germs or microbes, whose life cycle needed to be studied. For instance, early discoveries definitively linked bacteria to anthrax and tuberculosis. Following this, public health experts laid emphasis on the need to isolate germs, examine them under microscopes, and find out how they spread from one person or animal to another, and from one country to another. These findings supported the viability of quarantines as a useful measure, although not the only one available to health authorities.

This situation colored the politics of the 1883 epidemic in Egypt. Immediately after its outbreak, the International Board of Maritime Quarantine established a medical commission to investigate the disease. The commission's goal was to trace "patient zero," the first recorded case within a specific population, and find out how they got sick. Composed of Ahmad Shafi'i, the board's Egyptian delegate at Jeddah, Salvatore Ferrari, director of the board at Damietta, and James Grant, chief medical officer of the Railways Department, the experts supported a theory of "importation": that cholera was imported from India—where they believed it was endemic—via the Suez Canal. Despite previous cholera outbreaks in Egypt (1831, 1848, 1850, 1855, 1865), they maintained that all of these were cases of importation from India, most likely through the Islamic pilgrimage routes to Mecca.[67] Salim Pasha, of the Cairo Board of Health, also supported the importation theory. Meanwhile, rumors on the streets were that cholera had been introduced by the British and that the disinfectants, remedies, and even food provided by sanitary authorities were poisoned.[68]

British officials categorically rejected importation theory, since it implicated British-controlled India as well as the Suez Canal shipping business, of which they controlled roughly 70 percent. Following the canal's construction, European states had pressured Britain to enforce strict controls on ships departing from India and to implement quarantine measures at Suez. The British, however, interpreted this through the lens of imperial competition. To them, quarantine measures were excessively disruptive, costing pointless losses to commercial trade. Instead of importation, British doctors argued that cholera was endemic to nowhere in particular. Invoking the principles of the miasmatic theory of disease, they argued that cholera could spontaneously originate from poor sanitation, rotting organic material, and polluted soils anywhere, including in Europe.[69] In other words, no germ needed to be imported from India for cholera to strike.

Politicized on both sides, cholera sparked scientific competition among empires. As the disease was becoming less prevalent in Europe due to improved sanitation, European epidemiologists were increasingly forced to chase it in the colonies. In August 1883, a commission organized under the auspices of the leading French bacteriologist Louis Pasteur arrived to Egypt to investigate the epidemic. However, one member contracted the disease, forcing the commission to return to France without definitive conclusions. Shortly after, the German imperial government dispatched a scientific expedition headed by none other than Robert Koch. In Alexandria, Koch set up a laboratory where he isolated a microorganism—the *comma bacillus*—that seemed to be associated with cholera. However, he was unable to prove direct causation to the disease.[70]

As the cholera epidemic waned in Egypt, Koch continued his research in Calcutta, India, where he reconstructed a small outbreak and successfully established the connection between the microorganism he had first isolated in Egypt and the symptoms of cholera. Koch discovered that cholera is transmitted through drinking water contaminated by the dejecta of cholera patients. This groundbreaking discovery had profound implications for public hygiene and civil engineering, making scientific control over water supply and sanitary waste disposal essential public health measures. However, it took another decade for some medical authorities in England and Germany to fully accept Koch's findings, and several more decades for the miasmatic theory of disease to fully fade from public understanding.

In late 1883, Evelyn Baring—later known as Lord Cromer—became Egypt's Consul General and the de facto ruler of Egypt, a position he held until 1907. Under his direction, the colonial regime reorganized Egypt's public health administration to suppress dissent. Baring pushed for restructuring the International Board of Maritime Quarantine to diminish its international character, aiming to give Britain more control over quarantine measures in the Suez Canal.[71] Doctors who supported the importation theory were demoted or removed from their position.[72] In early 1884, the Cairo Board of Health was reorganized as the Department of Public Health and Sanitary Services, and staffed with several British doctors and engineers, including doctor Fleming Mant Sandwith, who became director shortly afterwards.

Later in that year, the government implemented some public health measures, including a controversial plan to replace mosques' traditional ablution basins with water taps. The plan was put forth by two British engineers: Colin Scott-Moncrieff, from the Public Works Ministry, and John Price, from the Sanitary Department. As the government tried to install faucets inside the popular mosque of Sayyida Zaynab, apparently without consulting religious leaders, it was met with widespread opposition. The rush to intervene in such a sacred space aroused suspicion, especially with the cholera epidemic still fresh in people's minds. Many questioned the permissibility of using the company's piped water for religious ablution, and the controversy led to the involvement of the Grand Mufti, who was asked to issue a judgment.

Given the novelty of the problem, the Mufti needed to look into the question. Sunni Islam, which predominates in Egypt, has four main schools of jurisprudence: Hanafi, Maliki, Shafi'i, and Hanbali. These schools, named after four classical jurists from the ninth to twelfth centuries, generally agree on core principles but differ on specific details, allowing for legal interpretation. After careful consideration, the Mufti ruled that faucets were permissible under the Hanafi school. Consequently, faucets became known in the colloquial dialect as *hanafiyya*, literally meaning in accordance with the Hanafi school.[73] Colonial observers cited this conflict as evidence of Egyptians' traditionalism and resistance to "godless improvements."[74] However, the conflict's rapid conclusion suggests that faucets quickly gained legitimacy, not just for religious ablution but for secular purposes as well. In fact, the initial rejection of the new technology cannot be separated from the political context of its introduction.

Unlike the Sanitary Department, Cromer reorganized government ministries according to the so-called "veiled protectorate," where the administration was ostensibly headed by Egyptians, but real power was vested in British advisors answerable directly to him. During that time, "Britons were imported like a flood into the Government service," as British irrigation engineer William Willcocks recalled decades later.[75] The "veiled protectorate" was a thin veneer coving the fact that Egypt had become a British colony, although still nominally part of the Ottoman Empire. In 1885, Scott-Moncrieff became the advisor to the Public Works Ministry, a position he held until 1892. In his diaries, Scott-Moncrieff recounts a conversation with Prime Minister Nubar Pasha that illustrates how the British takeover was marked by casual condescension toward Egyptian employees:

> As I have stated, Nubar on taking office sent for [Edgar] Vincent, [Clifford] Lloyd, and myself. He explained to us that the ministers, heads of departments, must be Egyptian subjects. It was against the policy of the British Government to appoint us three as actual ministers. Then he said he wished to speak to us each separately. What he said to Vincent and Lloyd I do not know, but his conversation with me was something as follows: After repeating that he could not make me Minister of Public Works, and that he must put in a native, he said: "*Voulez-vous avoir un homme capable, ou une nullité?*" [Do you want to have a capable man, or a worthless person?] "*Une nullité, s'il vous plait, Excellence,*" [A worthless person, please, Your Excellency] I replied. "*Ah, mon cher, vous avez raison, vous avez raison, je vous chercherai une nullité*" [Ah, my dear, you are right, you are right, I will find you a worthless person]. And he was as good as his word, and appointed a very nice old fellow—['Abd al-Rahman] Rushdi Pasha—to be my nominal chief.[76]

The Cairo Sanitary Commission

Two years after the cholera epidemic, the British colonial government established a special sanitary commission with the stated objectives of preventing future outbreaks and mitigating their destructive effects. The 1885 Commission d'assainissement du Caire (Cairo Sanitary Commission) met ten times to discuss the city's most pressing public health problems. Presiding over the commission was the German doctor Franz Engel, director of medical statistics and sanitary consultant at the Ministry of Charitable Endowments, which managed mosques, public baths, and water fountains. Other members included Egyptian doctor Ahmad Shafi'i, who had investigated cholera in Damietta two years earlier; French engineer Julien Barois, general secretary of the Public Works Ministry; Latif Salim, sub-Governor of Cairo; Jean-Baptiste Pierre, director of the Cairo Water Company; and Ayerst Henham Hooker, a British doctor at the Sanitary Department.[77]

The commission zeroed in on Cairo's traditional water and sanitation systems, identifying several major issues, including the Khalij Canal, mosques, and waste disposal. The Khalij Canal had been a focal point for urban reformers since the 1840s.[78] The commission found out that the canal was used as both a water source and a waste collector: seventy-seven cisterns extracted approximately 20,000 cubic meters of water annually from the canal, while 503 drainage pipes from nearby mosques and homes discharged waste into it.[79] Ideally, the drains were closed year-round, except for a brief period before the canal's opening. People were allowed to draw water again only after several days of running water, when the canal was deemed sufficiently clean. As it turned out, however, this arrangement was not always followed. The commission discussed implementing regulations, inspections, and fines to address the problem, but concluded that monitoring a four-kilometer canal running through dense urban areas was too challenging. In its final report, the commission recommended filling in the canal and converting it into a street with an underground sewer line.

This recommendation overlooked the dissenting opinion of Salim, who advocated for preserving the Khalij Canal.[80] Salim had argued that eliminating the Khalij would deprive several communities of their "right" to access water, adding that the government would need to compensate mosques, fountains, public baths, and private homes for the water they freely obtained. Rather than converting the Khalij into an irregular and "unsightly" street, he proposed regulations as well as reliance on police and local vigilantes for enforcement. Undeterred, Engel dismissed Salim's concerns, maintaining that the company's standpipes would suffice. However, the plan to fill in the Khalij was soon abandoned due to fears of popular backlash.[81]

The commission identified mosques as another significant public health issue, giving special attention to ablution courts and cesspits. It recommended installing faucets in all new mosques. With regard to existing mosques, the commission proposed prioritizing major ones like al-Azhar and al-Husayn, mandating that cesspools be emptied at least once a year. It also suggested limiting the use of mosque latrines and advocated instead for the construction of public toilets under the supervision of the Sanitary Department and the police.[82] Mindful of the Sayyida Zaynab Mosque incident, the commission recommended close coordination with the Ministry of Charitable Endowments in implementing these measures. In the following years, faucets were installed in over ninety mosques.

The commission classified percolating cesspools as a significant public health hazard. Still influenced by the miasmatic theory, the members expressed concern about soil contamination and its adverse effects on health. They could not agree more with Onofrio Abbate—an Italian doctor and long-time Cairo resident—who had described percolating cesspools as "a permanent abode of pestilence underneath one's house."[83] In their opinion, the city's hygienic conditions were deplorable even in affluent areas due to cesspool construction.[84] The commission recommended making all cesspools waterproof and ensuring regular emptying. However, they recognized that this would necessitate a more thorough upgrade of waste management methods, since waterproof cesspools would fill up more quickly and could not be managed by the current waste carriers.[85] Hooker, for instance, proposed introducing tank carriages, similar to those used in European cities. Equipped with air pumps, these tankers could remove cesspool contents with minimal human intervention. He also said that the new service could be managed by the government or a private company. The commission supported this solution, noting that the government had already received several business proposals along these lines. Although the concept of a city-wide sewage network was revisited, it was again quickly dismissed due to the anticipated cost and concerns about overwhelming residents with too many "innovations."

The commission also examined the five sewers constructed a decade earlier and initially recommended extending them to other parts of the city.[86] Two years later, however, this decision was reversed following a typhoid fever outbreak. When the Sanitary Department inspected these sewers, it uncovered "horrifying" conditions. Inspectors found that surrounding homeowners had been connecting their toilets to these sewers, contrary to their original design and intended use as surface water drains. Moreover, the sewers had been constructed with porous bricks and laid without sufficient incline, conditions that led to frequent clogs and the accumulation of dangerous gases. In fact, three workers died of asphyxiation while attempting to clear a sewer line.[87]

Finally, the commission reviewed two recent decrees related to housing and hazardous industries. In 1883, the government had issued a Ministerial Decree on Workmen's Dwellings, establishing sanitary standards for poor housing.[88] However, the commission found that the decree had not curbed the spread of substandard housing, particularly in Bulaq and Old Cairo.[89] In addition, the commission examined a decree categorizing dangerous establishments based on risk levels. The first category included slaughterhouses, tanneries, salted fish processing, quarries, iron foundries, petrol depots, and facilities producing chemicals, dyes, alcohol, matches, bricks, glass, fertilizers, dung cake fuel, and gunpowder. The second category encompassed public baths, laundries, public kitchens, distilleries, animal stables, food markets, grain and oil mills, meat curers, charcoal depots, and manufacturers of beer, paper, and soap. As opposed to second-category establishments, those in the first category required a government permit to ensure a minimum distance from residential areas. The commission emphasized that while dangerous industries should be moved away from residential areas, they should remain reasonably close to where workers live. Ultimately, many of these industries were relocated to Bulaq, Old Cairo, and 'Abbasiyya.[90]

In 1886, the Cairo Sewage Transport Company was established by French and British investors to upgrade waste disposal.[91] Following Hooker's recommendations, the company introduced a fleet of specialized tankers for clearing cesspools (Figure 9). Authorized by the government, the company emptied the cesspools of administrative buildings, public latrines, mosques, train stations, hospitals, prisons, and private homes, but only if the cesspools were waterproof. The company also acquired land in Old Cairo to build a factory for converting sewage into organic

FIGURE 9. A tank carriage belonging to the Cairo Sewage Transport Company. *The Power Wagon*, January 1908, 13.

manure. Although some hailed this new service as "sensational progress," its impact was limited.[92] The company's clientele remained a small fraction of Cairo's population, while the majority continued to depend on percolating cesspools and traditional waste carriers.

The Water Company Secures its Future

Once regarded as a luxury, access to abundant clean water was becoming an essential public health necessity by the closing decade of the nineteenth century. Shifts in medical theory and public health policy created new opportunities for the Cairo Water Company. Between 1887 and 1889, the company capitalized on these developments, advancing its position through two agreements with the Egyptian government. While these agreements resulted in lower water rates, they also reinforced the concession's legal standing and secured the company a larger share of future profits.

Even though the company's concession was valid for ninety-nine years, Article 13 of the concession contract granted the government the option to purchase the company after twenty-five years, calculated from 1870, when water supply actually began. In 1887, with the redemption period still several years from its scheduled end, the company and the government negotiated a new agreement: the government postponed its right of redemption until 1938 in exchange for some adjustments. One of these was a reduction in the water rate from one franc (thirty-eight millimes) per cubic meter to 0.7 franc (twenty-seven millimes).[93] The other was a profit-sharing arrangement. The government stood to gain 50 percent of future net profits between 1,350,000 and 1,750,000 francs, and 75 percent of net profits above that threshold.

These terms need to be viewed in context. Although the reduction in the water rate was substantial, the new rate remained quite high, not only for Cairo's residents but even by European standards. It was nearly double the current rate in Paris—0.4 franc per one cubic meter.[94] Comparing with English towns is more complex, as water rates were based on rental values, but rough estimates suggest that Cairo's rate was several times higher.[95] As a government insider noted in 1895, "the tariff of the Cairo Water Company would no doubt rank as a remediable point in the estimation of most of the ratepayers, seeing that it is at least ten times higher than that of any similar establishment in Europe."[96] Whereas water rates in European cities continued to decrease due to political pressure on private suppliers, Cairo's rate remained unchanged from 1887 to 1938. Additionally, the profit-sharing terms were not particularly advantageous, given that the company's net profits in 1887 were 1,353,000 francs, just above the threshold for the 50 percent profit-sharing bracket.[97] In effect, the company did not anticipate significant profit losses in the short term.

When the shareholders convened to ratify the agreement, many voiced concerns but ultimately approved it due to the benefits it offered the company. Jules Pastré, one of the original cofounders, remarked that the 75 percent profit-sharing clause was a "great sacrifice," but maintaining favorable relations with the government was clearly more important.[98] Seffer Pasha, a Pole who had joined the Ottoman army before settling in Egypt, also objected to the government's profit-sharing conditions. As an owner of 3,000 shares, he would have preferred either an annual fixed sum or a water allowance. Other board members present included Maurice Maunoury, a French politician and investor with a gas concession in Port Said; Charles Beyerlé, a French businessman and board member of the Crédit Foncier Égyptien (Egyptian Mortgage Bank); and Félix Paponot, a French contractor who had been involved in the construction of the

Suez Canal.[99] In the end, the shareholder ratified the agreement, and the company issued a new fee schedule detailing rates and subscription methods.[100]

In 1889, the company and the government signed a new agreement in which the government relinquished its share of profits and the right to redeem the concession in exchange for 4,000 of the company's dividend-paying shares.[101] At first glance, the new agreement seemed less favorable to the company, as one of the shareholders noted during an extraordinary board meeting. However, viewed over the long term, the 1889 agreement significantly strengthened the company's position. Beyerlé—the president of the board of directors—emphasized that the company was still in its early stages. If future profits significantly exceeded current levels, as he hoped, the new agreement would secure a larger share to the company.[102] Over the following decades, Beyerlé's expectations were largely realized, and the government lost a substantial share of profits as a result. As will be discussed in Chapter 6, the 1889 agreement was later criticized harshly for enabling this loss of revenue.

Conclusion

With the support of Nubar Pasha, the Cairo Water Company became one of Egypt's most securely established business monopolies. Within two decades of its founding, the company profoundly transformed the built environment and everyday life. More than just a service provider, the company played a pivotal role in Khedive Isma'il's urban development program, actively participating in the construction of streets and shaping their layout. Until the cholera epidemic of 1883, the company's presence in the old city was limited to public standpipes. Shortly thereafter, the company began supplying mosques and other institutions that had previously depended on water carriers, wells, and cisterns.

But this was only the beginning, since the vast majority of the city's homes were still not connected to the company's network. Two key developments are crucial for understanding how most building eventually became connected. The first is the building boom of 1897–1907 (Chapter 4), during which the built environment was redesigned to accommodate increased domestic water usage. The second is the construction of the sewage system in 1915 (Chapter 5). As I argued in the introduction, it was mechanized drainage that ultimately completed the transition towards domestic water consumption as a global norm. Thus, to fully grasp the revolution in water consumption, we must turn to the history of urban drainage. The next two chapters will examine the government's early attempts to build a sewage system and consider the lessons learned from its failures.

TWO

When Engineers Collide

Designing the Infrastructure of Private Water Consumption

Toward the end of 1889, the Egyptian government reopened the project to build Cairo's sewage system. The idea had surfaced a few times in the recent past but was quickly dismissed due to the anticipated high cost and the logistical challenges of implementing a city-wide infrastructure designed for daily public use. In the wake of the 1883 cholera epidemic, scientific breakthroughs in understanding the link between disease and sanitation helped prompt renewed interest on the part of the British colonial government. At the same time, Egypt's financial situation had slightly improved compared to the early years of British occupation, which had been marked by what Alfred Milner, a British colonial administrator, referred to as "the race against bankruptcy."[1] Since the occupation came on the heels of state bankruptcy, one of the immediate priorities for Egypt's colonial rulers was to balance the budget and ensure repayment to European creditors. Therefore, the colonial regime significantly curtailed public expenditure, except for irrigation, a cornerstone of colonial policy. By 1889, however, state revenues began to recover, allowing the administration to cautiously reconsider its economic plans.[2]

In this context, the colonial administration hired Baldwin Latham, a British civil engineer, to design Cairo's sewage system. Latham was a distinguished expert and author of influential manuals on water supply and sanitary engineering. After studying Cairo, he proposed a system that came at the hefty price of L.E. 550,000. To put this figure in perspective, the annal expenditure for all public works in Egypt between 1882 and 1892 averaged L.E. 772,500.[3] In fact, this project would have been one of the largest engineering undertakings by the colonial regime, comparable only to the complex repair of the Delta Barrage, which cost close to one million pounds between 1885 and 1890.[4] Given the substantial cost, the government convened a commission of experts to review the project and conduct a thorough cost-benefit analysis. The commission included three British engineers: Colin Scott-Moncrieff, the "advisor" of the Public Works Ministry and president of the commission; John Price, from the Department of Public Health and Sanitary Services; and A. C. Reid, director of the Delta Barrage.[5]

One might assume that Latham and the three other engineers would have a smooth working relationship. After all, they served the same empire. However, this was far from the case. The engineers collided over fundamental aspects of design. Reid, representing the government's perspective, argued that the scheme was too elaborate and "founded on a model applicable to an European town," while Cairo was unique in being "partly European, partly Oriental."[6] Scott-Moncrieff criticized the project's core features as "very expensive" and "of no use."[7] In response, Latham accused Reid of being "anxious to study economy [i.e., cost reduction] before all things," and therefore unable to appreciate the technical and public health merits of the proposal.[8] This chapter shows that this collision resulted from a fundamental disagreement regarding how to design technical systems that provide essential services to the public. The government experts wanted to tailor the design to reduce costs, making the project more palatable to their superiors. Latham, on the other hand, viewed it as a mistake to sacrifice efficiency for the sake of economy, as it defeated the purpose of building a safe and useful infrastructure in the first place.

With the march of industrialization, however, soil degradation became widespread in Britain, continental Europe, and North America. As Foster notes, the 1840s witnessed "a phenomenal increase in the demand for fertilizer."[13] So desperate were European farmers that they scavenged battlefields for human bones to use as fertilizer.[14] The crisis prompted European countries to compete for fertilizers thousands of miles away, particularly off the cost of Peru, where islands with large deposits of guano (bird droppings) were located. There, the abundance of fish and arid climate contributed to the formation of a nutrient-rich, dry fertilizer. The first commercial shipments of Peruvian guano arrived in Europe in the 1830s. Two decades later, a "guano mania" was in full swing, with thousands of tons of the precious material imported from overseas. By the 1860s, however these guano deposits were largely exhausted.[15]

It was in this context that European scientists and intellectuals began to take the idea of recycling human waste seriously.[16] French socialist philosopher and political economist Pierre Leroux proposed the idea of the "circulus"—a virtuous cycle between the city and the countryside, where humans would return to the land what they had taken from it.[17] Leroux believed that using human waste as fertilizer could significantly boost agricultural production and avert the Malthusian resource trap. In this way, recycling could complement wage labor as a source of value.

German chemist Justus von Liebig arrived at similar conclusions but via empirical routes. Liebig was one of the first to warn that European soil was becoming depleted due to mechanized farming methods and increasing demand for food production, describing the relationship between industrializing cities and the countryside as a "robbery system," where mineral resources were extracted from the soil without returning anything in kind.[18] His 1840 book *Organic Chemistry* clarified the role of essential nutrients like nitrogen, phosphorous, and potassium in plant growth for the first time. In his later works, Liebig continued to highlight the ecological consequences of this robbery system, arguing that soil productivity was declining at a rate that the existing supply of fertilizers could not remedy. Bypassing this bottleneck, he suggested, would require discovering "deposits of manure or guano . . . in volumes approximating to those of the English coalfields."[19]

And yet, the solution was readily available everywhere in Europe. Liebig argued that recycling human waste as fertilizer could not only avert the crisis but also replenish the soil for future generations. Therefore, he criticized waste disposal methods where toilet refuse was simply flushed into the nearest river. He wrote: "The introduction of water-closets into most parts of England results in the irrecoverable loss of the materials capable of producing food for three and a half million people every year."[20] For Liebig, rethinking urban sanitation was necessary for overcoming the material limits of nature under the conditions of modern agriculture.

Liebig's ideas profoundly influenced a generation of writers, including Victor Hugo and Karl Marx. Hugo approvingly cited Liebig's research in *Les Misérables*—in a section titled "The Land Impoverished by the Sea." There, he lambasts the absurdity of European nations sending fleets halfway around the globe to collect bird droppings while discarding their cities' sewage into the sea, thereby wasting an even better fertilizer readily available.[21] As for Marx, his engagement with Liebig was crucial in developing his ecological critique of capitalism, as several studies have shown.[22] Marx shared Liebig's view on the irrationality of contemporary waste disposal methods from an agricultural standpoint: "Liebig rightly criticizes the senseless wastefulness which robs the Thames of its purity and the English soil of its manure."[23]

However, Marx placed Liebig's analysis of modern agriculture in the larger context of industrial capitalism. He argued that the relentless pursuit of capital accumulation created an "irreparable rift" in the metabolic relationship between humans and nature, impoverishing both society and the environment.[24] As he states in Volume I of *Capital*:

> Capitalist production collects the population together in great centres, and causes the urban population to achieve an ever-growing preponderance. This has two results. On the one hand it concentrates the historical motive power of society; on the other hand, it disturbs the metabolic interaction between man and the earth, i.e., it prevents the return to the soil of its constituent elements consumed by man in the form of food and clothing; hence it hinders the operation of the eternal natural condition for the lasting fertility of the soil . . . All progress in capitalist agriculture is a progress in the art, not only of

> robbing the worker, but of robbing the soil; all progress in increasing the fertility of the soil for a given time is a progress towards ruining the more long-lasting sources of that fertility. . . . Capitalist production, therefore, only develops the techniques and the degree of combination of the social process of production by simultaneously undermining the original sources of all wealth—the soil and the worker.[25]

While Thomas Malthus believed in an inevitable resource crisis caused by population growth, Marx argued that capitalism was the root cause of this crisis, having set in motion dynamics that increasingly severed the relationship between humanity and the environment.

The technical development of mechanized sewerage in the second half of the nineteenth century unfolded against this backdrop. As growing anxieties over soil fertility led to a relentless search for fertilizers, and as guano islands were depleted within a few decades, the idea of converting human waste into fertilizer gained traction. However, the feasibility of implementing this idea on a city-wide scale remained uncertain. As will be shown next, engineers and businessmen were aware of these circumstances. In designing "air suction" systems of sewerage, they sought to meet the demands of hygiene and convenience while also turning human waste into profits. However, they first had to outcompete the "water carriage" system, which was beginning to establish a reputation for being simple, hygienic, and economical relative to other options.

Waterborne Versus Airborne Sewerage in Cairo

As the name suggests, water played a central role in the water carriage system, beginning from flushing waste down the toilet to ensuring that sewage was sufficiently diluted to flow easily through pipes. This system was first developed in Britain, with the London sewers—built in 1858—as the most successful and spectacular example. In contrast, airborne sewerage relied on compressed air for flushing and for moving contents through miles of pipes. Air suction systems were simultaneously developed by entrepreneurs in the United Kingdom and the Netherlands. The choice between these two systems sparked worldwide debates among engineers, doctors, and urban officials.[26]

Cairo was one of the cities where the two systems competed. However, it was never the intention of the British regime to initiate this competition. Initially, the government hired Baldwin Latham with no plan to compare his proposal to any other design. In fact, it was the idea of the Italian Doctor Onofrio Abbate, who was appointed by the government to review Latham's proposal. It is unclear whether Abbate had any material interests at stake. What is clear, however, is that he vehemently opposed waterborne drainage from a public health perspective. Still under the influence of miasmatic medical theories, Abbate considered it a "fatal mistake" to combine human excreta and household water and, worse, to channel them underneath the street.[27] Instead, Abbate urged the commission to consider an alternative "air suction" system, which the British engineers on the commission did consider, although without much enthusiasm, as will be seen.

Latham was one of the most prominent experts in waterborne sewerage. Born in 1836, he began his career as a surveyor to the town of Ely, near Cambridge.[28] In 1863, he moved to Croydon where he served as engineer and surveyor to the local board. During this period, he designed water supply and sewerage projects in several towns, including Croydon, Birmingham, Harrow, and Rugby. In 1871, he established a private consulting engineering business, but he remained close to policymakers through active participation on parliamentary committees and in professional societies, such as the Institution of Civil Engineers, the Royal Meteorological Society, the Royal Geographical Society, and the Royal Sanitary Institute. In 1865 he wrote *Papers Upon the Supply of Water to Towns* and in 1873 *Sanitary Engineering: A Guide to the Construction of Works of Sewerage and House Drainage*, which became a "sanitary classic," solidifying his reputation as a leading authority in the field.[29] This reputation paved the way for Latham's imperial career in 1889 when he was recruited by the Foreign Office in London. Over the next two years, he travelled between Egypt and India, studying and designing drainage systems for Cairo, Calcutta, Bombay, Benares, and Nainital. Thus, his career not only marked the consolidation of British sanitary engineering at home but also its international expansion by the century's end (Figure 10).

As a sanitary expert, Latham believed that improving Cairo's hygienic conditions required lots of water. Like other proponents of waterborne

FIGURE 10. Baldwin Latham playing cards. Courtesy of the Worshipful Company of Makers of Playing Cards, United Kingdom. As noted in Latham's obituary, "to those who knew Mr. Latham as an engineer and man of science only he seemed to be immersed in work of the sternest aspect; but he expanded to a surprising extent in historical studies. He revealed his civic associations to many of his friends by the gift of highly ornate packs of cards one Christmas, and greatly mystified them by the unexpected act until it was noticed that the ace of spades bore his familiar countenance, enshrined there as the Master of the Company of Playing Card Makers." "Obituary of Baldwin Latham (1837–1917)," *Symons's Meteorological Magazine* 52, no. 617 (1918): 50.

drainage, he believed that using large amounts of water served two key purposes. First, it aligned with emerging norms of public hygiene. Second, it minimized the risk of sewer clogging and system failure. As studies have shown, Edwin Chadwick, a notable English sanitary reformer, was perhaps the most influential advocate of the water carriage system, because it facilitated the use of abundant amounts of water, and therefore promoted public health.[30] In his opinion, the best response to the problems of urban growth and repeated cholera epidemics was to turn cities into machines for supplying copious amounts of pure water and for draining it away after consumption. British civil engineers, including Latham, developed the technology and the expertise to put these ideas into practice, and to adapt them to suit different cities.

The proposal Latham submitted to the Egyptian government in December 1889 was based on a set of simple recommendations: to build a drainage network and connect as many homes to it as possible, and to abolish the existing system of waste disposal based on carriers and percolating cesspools.[31] Examining the report provides a glimpse into the research Latham conducted before submitting the design. Having carefully studied Cairo's geography, he noted that the city was generally flat, which made it unsuitable for a system based solely on gravity. Accordingly, he designed a system that relied on mechanical pumping, dividing the city into twenty-eight drainage districts, each equipped with a small pumping station. These districts would be linked to a main collector running beneath the Khalij Canal and terminating at a central pumping station to the northeast of the city. Latham also recommended establishing a sewage farm where sludge could be used for agriculture—a precursor to today's treatment plants. He proposed placing the farm in the desert of 'Abbasiyya, then on Cairo's outskirts, maintaining that the farm's revenues could help cover ongoing maintenance costs.[32]

The design envisioned the rise of a new mode of water consumption centered on the home. As Latham stated in the report, only a small fraction of houses—2,600 out of 39,000—were connected to the Cairo Water Company.[33] However, he believed that the introduction of a sewage system would gradually change this situation, encouraging increased water usage at home. Nevertheless, Latham recognized that the transition would have to be gradual, as it depended on the pace at which houses were connected to water mains and public sewers. Over time, he predicted, reliance on mosques and public baths would diminish as the use of home toilets and baths expanded.

Abbate categorically rejected this proposal. He argued that the water carriage system had been viewed favorably due to its "collective simplicity," yet without adequately examining its hygienic implications.[34] The most significant flaw, in his view, was the potential contamination of the soil caused by leakage, which could vary considerably depending on the network's construction. Ultimately, these concerns were influenced by the miasmatic theory of disease, and specifically by the views of Max von Pettenkofer—the renowned German chemist and hygienist.[35] In fact, Abbate cited Pettenkofer's critique of the water carriage system: "It is harmful for public health, especially in cities where the soil is very porous."[36] For Abbate, Cairo's exceptionally porous soil was a prime example. The city, he said, stood on a "vast sponge" that could absorb large amounts of sewage, leading to unhealthy results. As an alternative, Abbate suggested exploring an air suction system called "Liernur," after its designer the Dutch engineer Charles Liernur.

Liernur began his career as a railroad engineer in the United States where he lived for two decades before the Civil War (1861–65). After the war, he returned to Europe where he began to develop a type of sewerage design using vacuum and compressed air. Instead of water closets, Liernur attempted to introduce "air closets" inside homes. The goal was to collect human waste "in its most valuable form—viz., fresh and undiluted," as a European commentator remarked without a shadow of sarcasm.[37] His vision was to create a city-wide system that collected human waste and converted it into dry compost with minimal human intervention, producing a limitless supply of fertilizer that was easily storable and transportable. While the water carriage system utilized sewage to irrigate farms in urban hinterlands, the Liernur system aimed to produce a dry fertilizer marketable worldwide. Liernur even claimed that manure sales would exceed the operational costs of sewerage, turning waste disposal into a profitable business. As guano reserves dwindled by the century's end, Liernur wanted to produce the human-based alternative. In fact, he was not alone in this pursuit; several other vacuum-powered systems, such as the "Shone system" and "Adams' sewage lift," were developed with the same goal of converting human waste into guano.[38]

However, the Liernur system was tested in several Dutch cities in the 1870s and 1880s with limited success. Experiments of varying scale were conducted in Amsterdam, Dordrecht, and Leiden, featuring vacuum-powered toilets connected to an underground sewer network in which air pressure was maintained by a central pumping station. This setup allowed human waste to travel several miles underground with minimal or no added water. Liernur also included an option for domestic water usage, provided that the water used was collected separately from toilet waste. He further envisioned recycling domestic and surface water for various mechanical operations.[39] Ultimately, however, Liernur was unable to scale the system up effectively, and his plan to produce human guano faltered.

Back in Cairo, the government commission shared Abbate's criticisms with Latham to solicit his response. As a proponent of the water carriage system, it is unsurprising that Latham held an unfavorable opinion of air suction systems, calling them inefficient and uneconomical. One major issue he flagged was the difficulty of maintaining proper air pressure within the pipelines. If water leakage posed a challenge in waterborne sewerage, he argued, air leakage posed an even greater challenge in airborne sewerage. Moreover, Latham stressed, his design was "as water-tight as possible," utilizing cast-iron sewers rather than the percolating brick and masonry sewers assumed by critics like Abbate. In addition, Latham argued that air suction systems came at a higher cost because they required constructing two separate networks—one for domestic water and another for toilet waste. According to him, the expense of this "double system of pipes" would far exceed the cost of his design.[40]

On this issue, the British engineers sitting on the government commission were in complete agreement with Latham. Signed by Scott-Moncrieff—head of the commission—a draft report reveals how they quietly dismissed Abbate's opinion:

> The commission have no means in Egypt of closely examining Mr. Liernur's system of sewage. ~~They are obliged to a large extent to be guided by the opinion of the expert specially selected by the~~

> ~~government to advise them. Independent of his opinion however, from what information they have been able to gather the Liernur system has not been a success in Holland~~. Mr. Latham considers it faulty in many particulars, expensive and complicated. ~~The commission accept this conclusion~~ [*sic*].[41]

In the end, the "collective simplicity" of gathering everything into the sewers prevailed. By the turn of the twentieth century, waterborne drainage had become the uncontested norm, especially in large urban areas. The spread of British engineering coincided with the expansion of the British Empire and facilitated this outcome.

Public Works Economics

To understand how Cairo's engineers examined the economics of the proposed sewage system, it is necessary to consider the broader context of how engineers around the world analyzed the economics of public works more generally. Economic analysis has been an indispensable component of engineering since at least the nineteenth century. With the proliferation of railways, bridges, and water systems, public works engineers were tasked with evaluating the costs and benefits of these infrastructures. This responsibility was especially important in light of the large initial investments required for these projects, a factor that both governments and corporations had to take into consideration. Governments and municipalities typically needed to justify these expenditures to the public and provide scientific evidence of the projects' benefits to society, whereas private corporations were primarily focused on generating profits for their investors.

By the 1840s, the field of engineering economic analysis had matured enough that some engineers began to draw general conclusions and share their findings with both specialists and the public. Charles Ellet, Jules Dupuit, and Dionysius Lardner are especially notable in this regard.[42] Their writings demonstrate how engineers studied markets generated by the construction of railways, bridges, and water systems, and how they identified the needs, means, preferences, and consumption patterns of prospective customers. Because of the novelty of their approach to questions of market demand, the field was later praised as contributing not only to "applied" economics but to economic theory itself.

Charles Ellet was an engineer from the United States. Known for his pioneering work on suspension bridges, he was also involved in railway and canal projects. He authored two important essays on the management of railways and canals on a sound economic basis: *An Essay on the Laws of Trade* (1939) and *The Laws of Trade Applied to the Determination of the Most Advantageous Fare for Passengers on Rail Roads* (1840). Writing as a corporate expert, Ellet underscored that the purpose of building such works was to "secure the greatest amount of revenue to the company."[43] To do this, economic managers must first estimate the "traffic" and the "value of trade" to be accommodated. How many passengers would use them on a regular basis? How much merchandise? In other words, good economic management begins with analyzing the market. Whereas engineers have long thought that they could only make an educated guess about market demand, Ellet argued that there was a "scientific" method to make these calculations. This method begins from the one element in absolute control of economic managers: the fees. A fee too high would turn away potential trade and passengers. A fee too low would increase the volume of trade and number of passengers but not necessarily generate the most revenue. Ellet theorized that there must be a point in the middle that produced the greatest amount of revenue, and proposed a mathematical formula for determining it.

In the railway essay, Ellet added considerations of "price discrimination" (also known as "market segmentation"): the practice of grouping customers into classes with different needs and means. For instance, he said that railway passengers should be divided into at least two classes, "the first consisting of those who regard cheapness as more important than the superior comforts and more select society offered in the best class of cars, and the second, of those who are willing to pay something for these considerations."[44] Therefore, he advised railway managers to design two different kinds of cars—one comfortable, the other less so—and adapt the ticket price accordingly. This method would ensure that each class of passengers paid an amount of money that was "proportional to their respective ability."[45] By dividing consumers into different groups, railway managers could generate more profits than by charging a single rate for all passengers. These ideas were later recognized by professional economists as offering a theory of monopoly, conceived as a profit-maximizing enterprise, albeit in a rudimentary form.[46]

Lardner's *Railway Economy* (1850) is perhaps the best-known text in the tradition of engineering economic analysis. Contributing to the book's fame was the fact that it was published in the wake of the British railway mania of 1846—arguably one of the most significant technology bubbles in history.[47] In the book, Lardner argued that railroads had far-reaching economic effects. For one, they cut transportation costs of merchandise, making goods cheaper. Moreover, the reduction in prices would generate broader demand effects: having saved some money, people may not only buy more of the same goods but also purchase items they previously could not afford.[48] In sum, Lardner contended that railways brought general economic benefits to society, advancing national wealth and prosperity. While he echoed some of the optimism of the railway mania, his conclusions were more measured, focusing on long-term effects rather than short-term gains. During the bubble, some railway experts claimed that replacing roads with railways would triple or quadruple the volume of traffic.[49] Though Lardner refrained from committing to a specific figure, he largely confirmed the idea that railways would increase the demand for transportation services and stimulate consumption across different groups of consumers.

In chapter 13 of *Railway Economy*, Lardner tackled the complex issue of railway charges. Determining the appropriate rate for merchandise and passengers' tickets was one of the most challenging tasks facing railway engineers. He proposed a mathematical approach to solve this problem, viewing revenues as the outcome of balancing two opposing forces. If charges were set too high, they would deter potential traffic; if set too low, they might attract more users but fail to maximize revenue. Lardner hypothesized the existence of an optimal point between these extremes at which receipts would be maximized.[50] The book featured a diagram illustrating this relationship as a bell curve, with charges as the independent variable and traffic as the dependent variable. The challenge for engineers was to visualize the curve for each proposed railway line and identify the point at which receipts would peak—the tipping point of the curve. This issue was not merely theoretical; it required collecting data and experimenting with small fare adjustments to observe the outcome.[51] Lardner's analysis was later praised not only for its practical relevance but also for its theoretical innovation, as it essentially introduced a mathematical model for understanding the relationship between price and demand for rail services.

If Lardner represents engineering economic analysis in Britain, Jules Dupuit epitomizes the French contribution to this tradition. Dupuit was a graduate of the École Polytechnique, which, together with the École des Ponts et Chaussées, gave France a reputation for distinguished engineering education. These schools trained engineers in mathematics and political economy to prepare them for public service.[52] As the French state emphasized scientific rationality and transparency in public finance, it directed its engineering corps to provide an objective assessment of the usefulness of public works to society. This information enabled the government to justify public borrowing and exercise eminent domain for potential land acquisition. In this context, state engineers developed a strong sense of professionalism and public duty, often putting them at odds with private enterprise.[53]

Dupuit, however, was different because his ideas aligned more closely with the French liberal school of political economy, which promoted laissez-faire capitalism, free markets, and individualism. These principles influenced Dupuit's approach to cost-benefit analysis. French state engineers typically operated on the assumption that public works were not built as profit-making enterprises. Therefore, they priced their services so as to cover the costs of maintenance and operation.[54] Moreover, they took the reduction of transportation costs as the primary indicator of benefits. For instance, suppose a railroad was built in place of a road. The conventional way to calculate the benefits was to add up all the money saved by passengers and merchandise through switching to rail transport. If these benefits exceeded the construction cost, then the railroad had net utility. If not, then there was no demand for it in the first place.

Dupuit gained his reputation by rejecting this conventional wisdom, viewing it as a misleading way of doing cost-benefit analysis.[55] He said that affordable, safe, and reliable transport lines would attract new customers who had previously been unable to travel. In addition, the significant reductions in transportation cost and prices of goods would free up money that could be invested elsewhere. Notably, Dupuit applied the same theoretical framework on examples of water supply, providing the following example: suppose there was a town located on high elevation where water supply was very costly. The introduction of more powerful pumps, for instance, would lead not only to reducing the cost of water transportation, but also to increased consumption of water. Customers who previously bought certain amounts of water to satisfy basic needs, could now buy more water to fulfill less urgent wants, such as cleaning houses and watering gardens more frequently.[56] In sum, Dupuit critiqued the existing formula of cost-benefit analysis for not being able to account for the full effects of new infrastructures. Such effects went well beyond cost reduction and shaped a higher demand for products and services.

Dupuit provided an alternative method for assessing the benefits of public works. This method hinged upon the ability of public works managers to gauge consumer demand, or how much the services are "valued" by consumers. In a remarkable overturn, Dupuit argued that the state should take lessons from businessmen. The businessman, he noted, "sets traps for the buyer's vanity and his credulity" to extract not just the cost of providing a service but also an additional amount reflecting what the consumer believes the service is worth.[57] For Dupuit, this mind game between provider and consumer offered a more accurate foundation for understanding the source of utility—not only for public works but for all goods and services. He emphasized that utility is variable, differing from one person to another and even for the same person under different circumstances. In essence, Dupuit argued that the "worth" of canals, bridges, and water systems was reflected in the monetary sacrifices the public was willing to make to use them. These sacrifices far exceeded the costs of production and transportation.

Initially met with indifference and skepticism, the ideas of Ellet, Dupuit, and Lardner were later celebrated for their significance to economic theory.[58] As historians of economic thought have shown, the 1870s marked a watershed moment in the history of the discipline, signaling a shift from political economy, with its traditional emphasis on production, to modern economics, which considered demand as the foundation of economic theory.[59] The leaders of this conceptual turn—economists like William Stanley Jevons, Léon Walras, and Carl Menger—specifically praised Ellet, Dupuit, and Lardner for their sophisticated understanding of economic demand.[60] In the final analysis, however, there remains a crucial difference between the two. Whereas the professional economists developed their ideas as a theoretical exercise, the engineers developed theirs through the process of designing, building, and managing infrastructures of transportation and water supply.

Pipe Economics: Analyzing the Economics of Cairo's Sewer System

As we saw, public works engineers had pioneered studying markets since the mid-nineteenth century, inventing new tools to analyze the needs and preferences of prospective consumers. What follows demonstrates that Cairo's engineers engaged in the same exercise, as they designed the city's drainage infrastructure. Cairo's engineers assessed the costs and benefits of drainage both on its own and in relation to water supply, which was controlled by the Cairo Water Company. As indicated earlier, the engineers involved in this project came to different conclusions. On one side was Latham, who designed the project with the assumption that the sewage system would transform current ways of living, leading to an increase in demand for private water consumption in the future. In contrast, the government's experts—Scott-Moncrieff, Reid, and Price—evaluated the project primarily through the lens of the costs to be borne by both the government and future users. In an effort to reduce the cost of the project, they proposed creating a two-tiered system: one for the newer quarters built after 1867—what they called "European Cairo"—and another for the old areas, what they labelled "Oriental Cairo." Had it been implemented, their project could have seriously inconvenienced most Cairenes.

One of the most critical steps in designing infrastructure is calculating its total working capacity: how many users will be served by a railway line or a bridge? How much electricity, water, or gas is the population projected to consume? The answers to these questions directly influence how engineers approach the design process, including that of sewage systems. Sanitary engineers begin by estimating the total volume of wastewater to be conveyed and then work backward to determine the specifications of the technical components. A system designed to handle the sewage of ten thousand people would require smaller pipes and less powerful steam engines than one serving a million inhabitants. Since pipes and pumps are typically the most expensive items in sewerage systems, their size and power have a direct effect on overall expenses.

Latham calculated the total capacity of Cairo's sewage system based on two types of water: domestic wastewater and surface water. The latter category included stormwater and infiltration, both of which had to be drained away for the purpose of convenience and public health.[61] He calculated surface water using data provided by the government, including maps, meteorological statistics, and graphs illustrating seasonal fluctuations of the water table. Calculating domestic wastewater, on the other hand, was more tricky because it required not only information on how much water people were currently using, but also an educated guess about how much they would use after the system was installed.

To calculate current water consumption, Latham obtained information from the water company on the number of houses and establishments connected to the water mains as well as the average volume of water distributed by the company's standpipes. He also examined the traditional system based on carriers, cisterns, wells, and public facilities.[62] As is the case with waterborne drainage in general, the volume of wastewater is directly linked to the total amount of water supplied to the city—with solids typically comprising only about 1 percent of volume.[63] Therefore, engineers typically estimated the volume of domestic sewage on the basis of water supply data.

When it came to future water consumption, it is unclear exactly how Latham made the estimates. Evidently, however, they were based on his experience with how water consumption had changed in other cities as a result of the construction of drainage networks. Latham reasoned that the new infrastructure would generate a higher demand for water services and calculated future water allowance accordingly. Another key factor to consider in this regard was the uneven geography of water distribution in the city. Only 2,600 houses out of 39,000 were connected to water mains, and they were concentrated in Azbakiyya, Isma'iliyya, and other new areas developed since 1867. As for the rest of the city, people continued to rely on the traditional system of public facilities, wells, and cisterns. To account for this disparity, Latham divided the city—then home to 414,000 people—into two classes of water consumers, assigning a different water allowance to each. The first class—100,000 people living in new areas—were given a water allowance of 136 liters (30 gallons) per day per capita. The second class—400,000 people living in old quarters—were assigned an allowance of 22½ liters (5 gallons) per day per capita.[64] Based on these estimates, Latham designed a system capable of conveying 65,500 cubic meters of wastewater daily.[65]

Based on this capacity, Latham estimated the total cost of the project at L.E. 550,000. The financial responsibility for connecting houses to public sewers was divided by property lines: the government would cover the cost for public areas, while homeowners would be responsible for expenses within their property. Recognizing the potential financial burden on users, Latham proposed the establishment of a government fund (of L.E. 100,000) to assist proprietors unable to afford connecting their homes. Additionally, he suggested that houses with very poor structural conditions could use manholes directly for sewage disposal.[66] The two most common methods of financing sewage systems were through direct government funding or sovereign debt. Assuming that the government would opt for the latter, Latham calculated that the debt could be amortized over a period of forty years, with an interest rate of 5 percent.

The three engineers on the review commission rejected key aspects of the proposal. First and foremost, they thought it was too expensive for Cairo, making it unpalatable to the colonial regime's financial controllers. Additionally, they argued that most households were too poor to afford using the system. Although the service did not involve direct fees, it imposed several indirect costs: connecting homes to public sewers, installing indoor plumbing and sanitary appliances, and obtaining enough water for flushing. These economic constraints, they contended, would exclude many prospective users. Reid drafted several notes in which he claimed that the system was "more elaborate than the existing conditions demand."[67] On this basis, he proceeded to cut down what he viewed as overblown or unnecessary components of the proposal. The first two to be eliminated were rain and infiltration waters, which Reid regarded as mere inconveniences rather than significant issues requiring a permanent solution.

Next, Reid cut down Latham's estimates for domestic wastewater. In the case of first-class water consumers—inhabitants of the so-called "European Cairo"—he lowered their allowance from 30 to 20 gallons per day per capita. As for the second class of water consumers—inhabitants of "Oriental Cairo"—Reid believed that Latham overestimated their consumption by a wide margin. Accordingly, he reduced their allowance from five to just one gallon per day per capita. With these modifications, Reid reduced the system's maximum capacity from 65,500 to approximately 11,000 cubic meters. As a result, Latham's proposed 9-inch pipes could be downsized to 6-inch pipes, and the 700-horsepower engine could be replaced with a 100-horsepower engine. This revision yielded a cheaper scheme that better suited the financial constraints of the colonial administration.

Reid had much to say about the second class of water consumers and why their water allowance should be slashed down. Citing the water company's statistics, Reid argued that the inhabitants of Cairo's old quarters bought an average of one gallon of water daily from public standpipes. This one-gallon figure, he maintained, was an expression of both means and needs. Because of the expensive water rate, people could not afford to buy more. At the same time, according to Reid, they did not need to: "The poor of an eastern country possess no boarded floors to be scrubbed, no table equipment to be washed daily, no changes of linen to be sent to the laundry, no elaborate meals requiring large quantities of water for cooking—they bathe only at a river, canal or a public bath."[68] If that was the case, then why, he asked, provide them with a five-gallon water allowance? In making these claims, Reid not only exaggerated the poverty of the old quarters, but also overlooked the substantial amounts of water sourced and distributed by traditional methods and used for various domestic purposes.

While Latham aimed to connect as many houses as possible and to install flushing toilets, Reid said that the city was not equipped for such a system. The latter contended that most of Cairo's residents were too poor to connect their homes to water supply, and therefore they did not have enough water to flush a home toilet. He warned that without sufficient water, the sewers would clog and release "dangerous gases." Additionally, he expressed skepticism about the ability of most Egyptians to use sanitary appliances correctly, particularly women, whom he believed could not be trusted with a flush toilet without education and supervision: "I believe it to be more practicable to educate the female population of the town to use the public latrines than to educate it into the use in a cleanly manner of private closets where no supervision or control is possible."[69] However, Reid left the door half-open, suggesting that if living standards improved sufficiently, private toilets could be introduced: "If in the future, the wealth and standard of comfort of the people rise sufficiently to establish a demand for private closets then they can be built and the necessary pipes laid down."[70]

Taking these factors into account, Reid proposed a two-tiered sewerage system: one for the "European" part of Cairo, and another for the "Oriental" part. In the first, all houses would be connected to water and sewers, making it possible for private bathrooms to be easily established. In the second part, however, Reid argued that no houses should be connected at all and that all existing private toilets should be abolished. Under this scheme, residents of the old quarters would be legally required to use only public facilities such as mosques, baths, and shared public latrines to be built and supervised by the government. This proposal meant that women, the sick, and the elderly would have been forced to leave the house and go to the nearest mosque or public latrine simply to use a toilet, regardless of day or night. Had it been implemented, this extraordinary and cruel idea would have imposed significant hardship on the population. Ultimately, it was a textbook example of colonial hubris and of technological design in the service of social, racial, and gendered hierarchies.

While Reid argued that living standards were too low to support Latham's vision of widespread domestic water usage, Price from the Sanitary Department sought statistical evidence to support this claim. In 1890, there were no official government statistics on living standards in Egypt.[71] Drawing on his experience as a sanitary engineer who regularly inspected buildings in old quarters and slum areas, Price decided to produce a rough statistic.[72] He reached out to the Municipal Bureau (Dayra Baladiyya), asking for data on the number of houses in Cairo and their rental values. He prepared the following table with the information he received:

TABLE 1. Distribution of Houses by Rental Value

	Houses
Paying rent exceeding L.E. 60 per annum	1245
from L.E. 36 to L.E. 60	1250
from L.E. 24 to L.E. 36	1812
from L.E. 18 to L.E. 24	1824
from L.E. 12 to L.E. 18	3098
from L.E. 6 to L.E. 12	6999
below L.E. 6	17263
Total	33,491
Huts[a]	5,634
Grand Total	39,125

[a]Slums were colloquially called *'eshash,* literally meaning huts.

Source: DWQ, Diwan al-Ashghal al-'Umumiyya (Public Works Ministry), file no. 4003-037344, Notes on Cairo drainage by John Price, April 1 (1890).

The table revealed that over half of Cairo's houses had a rental value of less than L.E. 6 per annum. Based on this data, Price estimated that 9,000 to 10,000 houses—about a quarter of all buildings—were in such poor condition that they required significant repairs before they could be connected to water mains or public sewers.[73] Furthermore, it was impossible to accurately estimate the cost of connecting these houses, as conditions varied widely and had to be assessed on a case-by-case basis. In Price's final analysis, most of Cairo's residents were simply too poor to install indoor plumbing or connect their properties to water and sewers.

Scott-Moncrieff, the highest-ranking engineer in the colonial administration, summarized the commission's findings in his final report. He concluded that the idea of connecting as many homes as possible was unsuitable for most residents, "since their habits and their poverty alike prevent their erecting in their houses water closets of an approved type."[74] Scott-Moncrieff accepted Price's assessment of living standards, estimating that half of Cairo's houses were in relatively poor condition. Connecting these houses to public sewers would be not only very expensive, but also useless if not actually dangerous "since it is certain that proper water closets could not be found in such houses."[75] Accordingly, Scott-Moncrieff endorsed Reid's two-tiered scheme, with the added modification of building 400–500 instead of 150–200 public latrines.

In March 1890, Latham received Reid's critical notes on his proposal, just as he was preparing to board the steamship Bokhara from India, where he had been studying various sanitary projects. During the journey, he composed a sixteen-page reply dismantling the government's position and the assumptions underpinning it.[76] To begin with, Latham acknowledged the poor condition of many buildings, but he said that the scheme was flexible, allowing for house connections to be made gradually as circumstances permitted. He emphasized that it was not necessary to connect all buildings at once; the system was designed to eventually connect every house "at some distant date," while residents in substandard housing could continue to use traditional public facilities in the interim.[77] Moreover, he rejected the "partly European, partly Oriental" argument. While acknowledging that most Cairenes used relatively low volumes of water at home, he argued that this did not preclude the possibility of private sanitation. He designed the sewers with "rapid falls" to ensure that the system could effectively transport solid waste even with minimal water usage.[78] This feature made the system particularly suitable for the second class of water consumers, as it accommodated traditional waterless toilets and eliminated the need for large volumes of flushing water.

Latham criticized the government's experts for underestimating current levels of water consumption. He noted that the one gallon of water Reid deemed sufficient was "entirely assumed in the face of the actual figures given in my report."[79] According to available statistics, the city consumed 1,140 cubic meters of water daily from standpipes and 700 cubic meters from the Nile and canals. Additionally, significant amounts of water were stored in cisterns and drawn from wells. Latham said that considering all these sources would "at once show the fallacy of the conclusions of the author of the notes [Reid]."[80] While it was impossible to determine exactly how much water Egyptians used daily, Latham demonstrated that one gallon was less than what they already consumed, let alone what they would want to consume in the future once the sewage system was built.

The most revealing part of Latham's long letter is where he addressed water consumption projections. Reid had assumed that water consumption would remain constant after the sewers were installed—that was why he thought the inhabitants of old quarters would continue using the same supposed one gallon of water daily. To Latham, Reid had failed to grasp the role of drainage as a catalyst for increased private water consumption. The main purpose of this new technology, he maintained, was to encourage greater use of water at home, both for public health and convenience. While acknowledging the existing constraints, Latham believed that waterborne drainage was very likely to change patterns of water consumption. Like the public works engineers discussed in the previous section, Latham emphasized the demand effects of the technology, arguing that the new infrastructure would transform the market.

As an expert on water systems, Latham was among the first to observe how cities interacted with new drainage networks and how patterns of water consumption evolved as a result. Drawing on his experience in India, he noted: "year by year with the introduction of sanitary works, the demand for water by an Oriental population are [*sic*] continually on the increase."[81] In the case of Bombay, "the entire population use at present sixteen gallons per head per day, and this quantity is found insufficient, and an additional quantity equal to seventeen gallons per head per day for one million people is now being brought into the city."[82] Therefore, the project had to be large enough to accommodate this probable future. Postponing the question would put the system under stress and lead to more expenses in the long run.

In the end, Latham underscored that the commission's single-minded focus on cost reduction had led them astray: "What does all this criticism of the author of the notes [Reid] mean, his own words supply the answer 'that an economy may perhaps be effected by reducing his figure'—that is the sewers are to be reduced in size for the sake not of efficiency but of economy."[83] To Latham, cost awareness was necessary, but making it the driving force behind engineering design was a grave mistake. It was tantamount to compromising the highest ideals of the profession to satisfy the employer. Therefore, in a private conversation with Scott-Moncrieff, Latham refused outright to reduce the size of pipes, declining to tailor the proposal to suit the budgetary constraints of the British administration.

By the late nineteenth century, engineers around the world were grappling with similar challenges. They sought to define a code of ethics guiding their work independently of the pressures of employers, whether governments or private businesses. Central to these struggles was the development of autonomous standards of technical efficiency, usability, and safety—issues that were debated within professional circles, often under public scrutiny.[84]

Imperial Competition

Towards the end of 1891, the project was shelved, but for reasons that are not directly related to the technical and economic discussions explored above. Indeed, the British regime was reluctant to finance the project because of the tight fiscal policies adopted since the beginning of the occupation. Commenting on the project, Elwin Palmer—the "advisor" of the Finance Ministry—recommended seeking a second opinion "before we incur such an enormous expenditure."[85] Lord Cromer, the British consul-general and de facto ruler of Egypt from 1883 to 1907, deferred the matter until the necessary funds could be secured.

A year later, French interference effectively ended what remained of the project. As the Egyptian government studied various funding options, it explored the idea of using the Octroi receipts—duties collected on consumer goods entering Cairo. However, these funds could not be used without France's approval. Following the political and economic crisis of 1876, a joint British-French Caisse de la Dette Publique (Public Debt Commission) was established to manage Egyptian finances. As a result, France held veto power over the use of any portion of Egypt's reserve funds.[86]

In May 1891, Tigrane Pasha, Egypt's Foreign Minister, formally requested the French government's approval to use Octroi receipts for funding the sewage system scheme. In a letter to the Marquis de Reverseaux, the French consul in Cairo, Tigrane emphasized the project's public health benefits, particularly for "the European colonies."[87] Reverseaux communicated the request to Alexandre Ribot, the French Minister of Foreign Affairs, noting that it would be difficult to refuse it given Cairo's unsanitary conditions. However, he suggested that raising property taxes might be a more appropriate solution than using Octroi receipts. This alternative, however, would require the Egyptian government to seek the approval not only of France but of several other foreign nations. According to the system of capitulations, foreigners living in Egypt and other Ottoman provinces were exempt from local taxation.[88] The London Convention of 1885 had allowed the Egyptian government to levy house taxes on foreign nationals for the first time, but increasing their value would necessitate another round of approvals from all the foreign Powers.[89]

It emerged that the French Minister of Foreign Affairs was aware of the sewage system project. In his correspondence with the French consul in Cairo, Ribot mentioned that "some talk about a figure above £. 600,000."[90] Other sources indicated that significant profits could be made from the transformation of the Khalij Canal into a boulevard with a commercial front, which would take place in parallel to extending a sewer line. Ribot argued that these profits should not be monopolized by English firms. He conditioned French approval on raising property taxes on the British regime's willingness to open the project to competitive bidding. Subsequently, Reverseaux facilitated an agreement between the French and Egyptian foreign ministers. France agreed to the tax increase provided that the Egyptian government held a public tender process in which French construction firms could compete for contracts. Thus, by the end of 1891, Latham's proposal became part of a large public tender file that included twenty-nine other proposals from engineers and businessmen from Britain, France, Italy, Germany, and Egypt.

Conclusion

Despite remaining on paper, exploring Latham's proposal and the extended discussions it led to among engineers and public health experts reveals significant insights into the hydrological conditions of Cairo, the potential obstacles facing the project, and its likely effects on future patterns of water consumption. While the views of government engineers were steeped in racialized assumptions, there was a grain of truth to the idea that Cairo's houses were ill-prepared for domestic use of large volumes of water. Up until then, most buildings were constructed with traditional materials like limestone and mud bricks, which made installing indoor plumbing and water fixtures technically challenging and potentially costly. As will be seen in Chapter 4, this situation began to change with the building boom of 1897–1907, which saw Cairo's residential architecture increasingly adapted for domestic water use. The construction of new types of housing began to remove one major obstacle facing domestic water consumption.

From today's perspective, it is hard to believe that waterborne drainage was once a novel technology subject to such intense debate. In fact, it was widely adopted and calibrated on a global scale only at the beginning of the twentieth century, with several alternative systems under consideration before then. As the above shows, the experts—or at least some of them—were cognizant of the potentially transformative effects of waterborne drainage on both built environments and everyday habits. If Latham's projection about increased demand for water services was an educated guess in 1890, it became a widely accepted principle in civil engineering by 1907, as will be seen in Chapter 5. At the same time, the significance of this principle was not lost on the Cairo Water Company. With its monopoly on water supply, it was uniquely positioned to benefit from the increasing demand for private water consumption that followed the installation of a sewer network. As will be seen in Chapter 6, the construction of the sewage system in 1915 ushered the company into a new era of profitability.

THREE

"An Insane Proposal"

Draining Cairo by Gravity Alone

The year 1892 marked a turning point for British rule in Egypt. During the first decade of the occupation (1882–92), colonial administrators focused primarily on avoiding bankruptcy and repaying the national debt. Nearly half of the country's revenues were transferred to European creditors, while the remainder was spent domestically, with irrigation taking precedence.[1] As the financial situation began to stabilize and as the British "occupation" became rather permanent, the colonial regime shifted toward more ambitious efforts to modernize the country and develop its agricultural, mineral, and financial resources.[2] Most notably, the government began exploring building a dam in Aswan, a project expected to expand cultivable land and significantly boost agricultural output, particularly cotton.[3] Around the same time, the colonial administration initiated several surveys aimed at gathering comprehensive information on Egypt's lands and mineral resources.[4]

In 1892, the government initiated a survey of state lands that gradually evolved into a nationwide survey of landed property.[5] Completed in 1907, this monumental undertaking produced thousands of detailed cadastral maps that were used as the basis for Egypt's new taxation system. As Timothy Mitchell argued in *Rule of Experts*, the turn to cartography signaled the emergence of a new form of power over space and resources.[6] By upgrading the taxation system through maps, the British regime significantly enhanced its capacity to extract the economic benefits generated in part by its irrigation program. In other words, British colonial development went hand in hand with building the bureaucratic apparatus capable of extracting economic surplus.

Also in 1892, the government produced a comprehensive survey of land titles in Cairo. Composed of more than 350 maps, the survey meticulously recorded the names of property owners, plot by plot, across the entire city (Figure 11). Like the national survey analyzed by Mitchell, the Cairo survey aimed to upgrade the taxation system through cartography. Whereas the old cadastral registers relied on written deeds, which verbally described the dimensions and boundaries of property, the survey produced a database of spatial information. Using these maps, the British administration planned to increase the house tax from 8.3 to 10 percent.

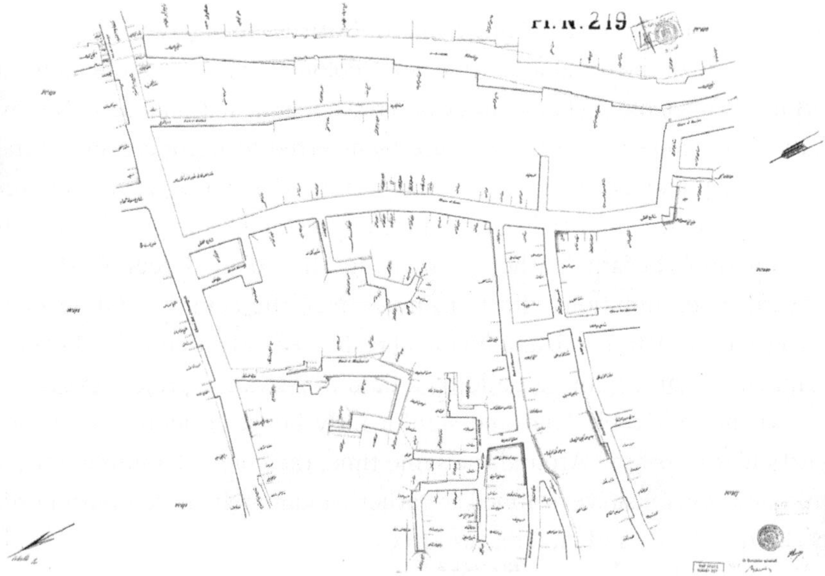

FIGURE 11. The 1892 Cairo survey, sheet no. 219 (Bab al-Sha'riyya). The Egyptian Survey Authority.

In essence, the 1892 Cairo survey anticipated the national survey conducted between 1892 and 1907.

Today, the 1892 Cairo survey is considered the single most important reference for property titles in historic Cairo. Housed at the Egyptian Survey Authority, black-and-white copies of individual map sheets are sold to the public for L.E. 1,000 each (in 2015), making it the most expensive historical map available for sale. These certified copies are used as evidence in court cases involving land ownership disputes. Typically, buyers are interested in one or two sheets—just enough to cover the plot in question. Due to its unique significance, the 1892 Cairo survey is closely guarded, with its large key map strategically displayed on the wall behind the director of the Map Sales Room. When I spent several weeks researching at the Survey Authority, I was intrigued to learn that it was commonly referred to as the "barawa" map and that no one knew the origin of this label—one senior official assumed it was some kind of serial code. In fact, it is an Arabized form of the proper name "Barois," referring to Julien Barois, the French engineer and General Secretary of the Public Works Ministry from 1884 to 1906. In 1892, he led a team of engineers and cartographers that produced the survey. What no one at the Survey Authority seemed to suspect was that the survey was originally created in the context of a sewage system project, with two interconnected purposes in mind: 1) to raise the house tax and use the added revenue to fund the project, and 2) to assist Barois in determining the optimal placement of sewer pipelines. In this way, the map now leads an afterlife entirely disconnected from its history in the sewers.

This chapter uncovers the story behind the 1892 Cairo survey and situates it in the broader context of the British regime's efforts to upgrade the bureaucracy and exert tighter control over the country's natural and financial resources. It examines the sewage system project for which the map was originally produced and analyzes why it ultimately fell through. Barois successfully convinced other experts of a sewage system scheme that, he said, was more elegant and cost-effective than the one proposed by Baldwin Latham two years earlier (Chapter 2). It was elegant because it proposed to drain the whole city solely by gravity, thereby eliminating the need for the small pumping stations that Latham believed were necessary. For the same reason, this scheme was expected to be less expensive, with an estimated cost of L.E. 500,000. However, when Barois conducted a more thorough study between 1892 and 1894, the project's estimated cost ballooned to over L.E. 1,000,000. The cost was so prohibitive that Lord Cromer, the British consul-general, postponed the project indefinitely. An anonymous government source later proclaimed that the attempt to drain Cairo by gravity alone was "an insane proposal" from the beginning.[7] To understand why the cost estimates doubled, we must delve deeper into Cairo's subsoil and the geology of the Nile Valley as a whole.

At the outset, it is worth noting several developments that unfolded in parallel. First, the project sparked considerable unrest in Matariyya—the now-popular neighborhood was then a distant suburb known for its fresh air and tourist attractions. Barois designated land adjacent to Matariyya as the site of the sewage farm, leading concerned residents to launch a "not-in-my-backyard" campaign, demanding government transparency and local participation in the project's planning and assessment. Second, as the government continued to rely on foreign expertise for large projects, Egyptian engineers grew more frustrated and began to organize themselves. The launch of *al-Muhandis* (The Engineer) journal in 1892 marks the emergence of an oppositional consciousness among Egyptian engineers under British rule. Finally, Cairo was struck by another cholera outbreak in 1895–96, which led to violent clashes between the police and students at al-Azhar University over the government's anti-cholera measures. After the epidemic subsided, the government decided to fill in the Khalij Canal, an idea that had been suggested repeatedly but abandoned due to the canal's popularity.

The 1892 International Commission

In 1892, the Egyptian government put the Cairo drainage project out to tender and appointed an international commission to oversee the process. The competition was strong, with thirty bids vying for lucrative government contracts. However, none of the offers met the minimal requirements, and the commission decided to draft its own proposal. Made up of three engineers from England, France, and Germany, the "international" commission was intended to be neutral, with a composition designed to facilitate judging the offers based on merits rather than nationality. While the commission's work appears to have been largely free of influence from the British colonial administration or broader imperial economic pressures, it was significantly swayed by one man: Julien Barois. After spending several weeks in Cairo, members of the commission returned to Europe, leaving Barois to finalize the details.

As seen in the previous chapter, the Egyptian government began experimenting in 1891 with the idea of raising Cairo's house tax to fund the city's drainage project. According to the system of capitulations, which granted extraterritorial privileges to foreigners living in Egypt, the government needed the approval of several European countries before levying taxes on their subjects. The most difficult one to obtain was that of France, Britain's most significant rival on the imperial world stage. As the British colonial regime realized, obtaining French approval would have to be in exchange for something else. In this case, it was in exchange for opening the Cairo drainage project to public tender, allowing French businessmen to compete freely for government contracts. In November 1891, the Egyptian government launched the tender process by placing advertisements in prominent British and French trade journals, including *The Builder, The Engineer,* and *Le Génie Civil,* as well as in *Journal Officiel,* the government's official publication:[8]

> MINISTRY OF PUBLIC WORKS
> The Government, intending to carry out the drainage of the city of Cairo, will receive, until January 31, 1892, all projects that will be sent to it for this purpose. The projects, with plans and supporting documents, if applicable, should be addressed to the Ministry of Public Works (Administrative Service). A premium of 200 pounds will be granted to the author of the project that will be judged the best. The Government reserves the right to accept any project deemed appropriate or even to accept none.[9]

Britain and France each appointed an engineer to oversee the bidding process and evaluate the technical and financial merits of the submitted proposals. A German engineer was included to serve as a tie-breaking vote if needed. Of the three selected jurors, the German James Hobrecht stood out as the most distinguished. Trained as a civil engineer, Hobrecht had expertise in both urban drainage and town planning. In 1862, he developed a large-scale extension plan for Berlin, later known as the "Hobrecht plan," which prepared the city for future growth and became a model for other German cities.[10] During the 1870s, Hobrecht focused on Berlin's drainage system, culminating in the publication of *Die Canalisation von Berlin* (The Sewerage of Berlin).[11] Beyond Berlin and Cairo, Hobrecht also advised on water and sanitation systems in Moscow and Tokyo. Meanwhile, France appointed Adolphe Guérard, Inspector General at the Ponts et Chaussées (School of Bridges and Roads), to the commission. Between the 1870s and 1900s, Guérard earned an international reputation as a port and harbor engineer, conducting projects in Istanbul, Salonica, Jaffa, Lisbon, Suez, and Panama.[12] The third juror, representing Britain, was the senior engineer Henry Law, author of introductory textbooks such as *The Rudiments of Civil Engineering*.

In the two months separating the advertisement and the submission deadline, several businessmen contacted the Public Works Ministry, requesting maps, censuses, and other information necessary to design a drainage scheme.[13] At the same time, they sought ways to gain an edge over their competitors. For instance, the French public works firm J. L. Michau & Fils submitted a proposal based on phased execution of the project. The plan was to first connect new neighborhoods and mosques, with the rest of Cairo integrated gradually as funds allowed.[14] However, Colin Scott-Moncrieff—the advisor to the Public Works Ministry—expressed skepticism about the plan, noting that partial implementation could be nearly as costly as sewering the entire city.[15]

In another instance, British engineer Isaac Shone energetically promoted his system to the Egyptian government.[16] Shone had been active in the sanitary engineering field for over a decade, having patented a type of pneumatic ejector in 1878, which was adopted by the British House of Parliament and several towns worldwide.[17] In a letter to the Public Works Ministry, Shone emphasized the advantages of his system, particularly its suitability for flat cities like Cairo. He also noted that his ejector would be featured at the upcoming World's Great Exposition in Chicago (1893). Shone requested an extension to the submission deadline to allow time for a more tailored proposal. However, his request was denied. While acknowledging the tight timeline, the ministry encouraged Shone to submit a draft proposal, even if incomplete, with the understanding that if selected by the international commission, he could later provide detailed studies and cost estimates. In January 1892, Shone traveled to Cairo and personally presented his proposal to Scott-Moncrieff, which included an overview of the Shone system, a brief note on its potential applicability to Cairo, and a list of towns where the system had been implemented, along with corresponding population sizes and costs.

Despite the short deadline, competition for the project was intense. By the end of January 1892, the Public Works Ministry had received numerous proposals from engineers and firms across a dozen countries.[18] The ministry compiled a substantial public tender file containing thirty offers, including previously submitted plans from Baldwin Latham, J. L. Michau & Fils, and J. E. Cornish from the Alexandria Water Company. Among the new submissions was an offer by John Price, sanitary engineer at the Egyptian Public Health Department, as well as several others by Egyptian engineers, including Mahmud Fahmi, an engineer at the Tanzim Department (Cairo City Services).[19]

The international commission convened several times in February and March of 1892. In one session, they established the selection criteria. According to them, an acceptable offer needed to include: a cost estimate, a rough execution plan, and an explanation of how the proposed system would improve Cairo's sanitary conditions. In another session, the commission met with the authors of proposals then present in Cairo. Despite these efforts, however, the three experts were ultimately dissatisfied with the outcome. Most offers, they said, consisted only of brief notes recommending generic schemes, often accompanied by incomplete sketches or drawings. Few addressed Cairo's specific needs, and even fewer provided sufficient technical and methodological justification. None of the submissions were based on a detailed map of Cairo or included essential information about the soil, streets, land levels, or existing water and gas pipelines. As the commission stated in its final report, "the most complete studies were preliminary drafts at best."[20] In the end, the three engineers concluded that none of the proposals fully met the criteria, with only three—those submitted by

J. L. Michau & Fils, Fahmi, and Price—being worthy of further consideration. However, rather than choosing between three incomplete drafts representing French, Egyptian, and British interests, the commission opted to exercise its right to draft its own proposal.

Judging from the available records, the international commission's deliberations were largely free of external influence, but with one major exception. While reviewing the offers, the jury took special interest in a note written months earlier by the French engineer Julien Barois.[21] In that note, Barois criticized Baldwin Latham's proposal (Chapter 2) as needlessly complicated, arguing that it was unnecessary to install several small pumping stations in Cairo's centrally located districts, as Latham had believed. According to Barois, it was feasible to design a gravity-based system where only the network's slope was enough to generate self-cleansing velocity. He argued that by eliminating these small pumping stations, the government would end up with a more efficient and economical scheme. The commission "entirely agreed" with this view, citing Barois in its final report.[22]

Convinced by the gravity-based approach, members of the international commission conducted some research before drafting their own proposal. For instance, they studied the findings of previous sanitary commissions, such as that of 1885 (Chapter 1), gathered data on water supply and soil conditions, and conducted field visits in different parts of the city. Within two months, they had completed a preliminary project, with an estimated cost of L.E. 500,000. Significantly, however, they confessed that this figure was tentative, as a more detailed study was required for a precise cost estimate. Since they anticipated the study to be painstaking and time-consuming, it was left to the Egyptian government to carry out this task.[23]

According to the international commission, Cairo should be divided into four drainage districts, each equipped with a main collector and a branching sewer network (Figure 12).[24] The collectors were to be installed at the greatest depths, ensuring a continuous downward flow of wastewater from homes to branch lines and, ultimately, into the collectors (Figure 13). The first collector was planned to run under the longest continuous street in historic Cairo, connecting Sayyida Nafisa in the south to Ghamra in the northeast. The second would be located

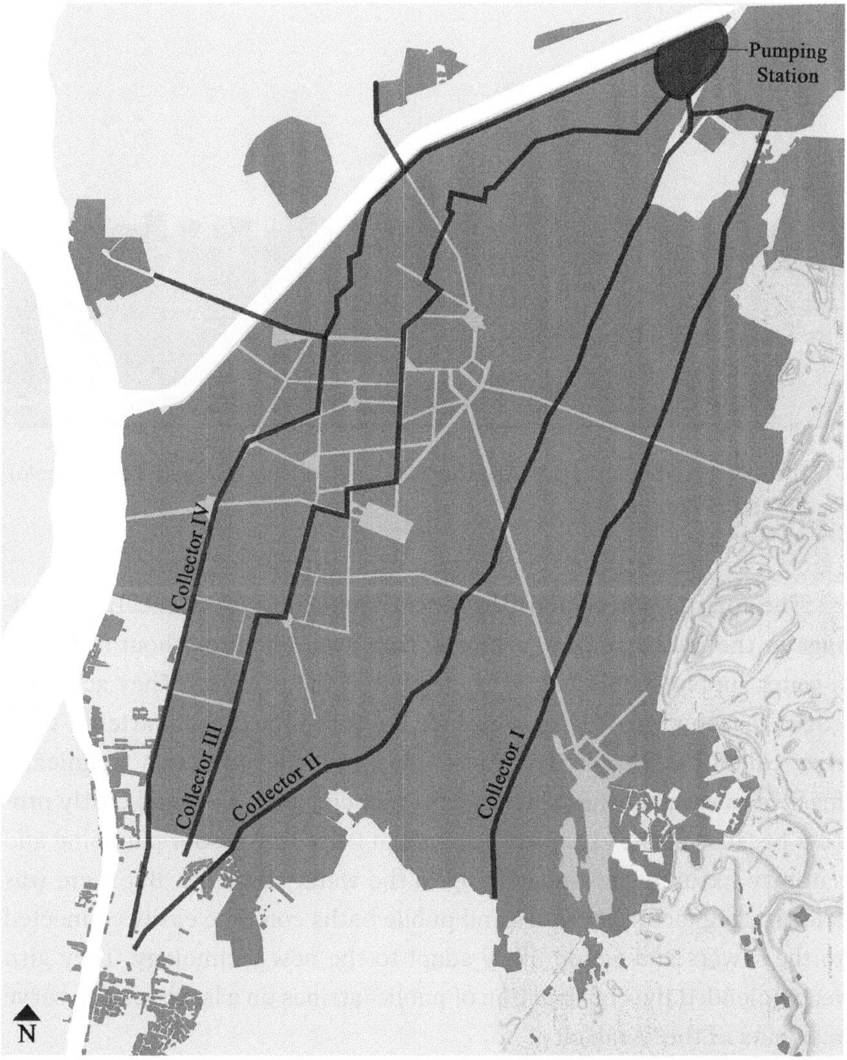

FIGURE 12. The 1892 Cairo drainage project. Illustration by Amr Essam.

under the Khalij Canal, while the third would start from the former site of gunpowder manufacturing and end in Ghamra. The final collector was planned to serve Bulaq and other low-lying areas near the Nile. Ghamra was chosen as an ideal site for a pumping station, from where sewage would be conveyed to a farm and used for irrigation and fertilization.

FIGURE 13. Longitudinal section of the 1892 Cairo drainage project (not to scale). Illustration by Amr Essam.

The commission established general guidelines for connecting buildings to the public sewers, with one firm rule: homes without access to a water supply should not be connected to the sewers.[25] They acknowledged that this would largely exclude houses in the older quarters—what they called the "Arab city." Although they believed it was technically feasible to connect these homes, they anticipated a slow and costly process because of the expenses involved in installing indoor plumbing and sanitary fixtures, and subscribing to the water company. But there was a mitigating factor: mosques and public baths could be easily connected to the sewers and would likely adapt to the new technology. They also recommended the construction of public latrines on a large scale to serve residents of the "Arab city."

To conclude, the international commission emphasized the need for a detailed study, which should include a highly detailed map of Cairo and a comprehensive land-level survey.[26] This information was essential to determine the optimal placement of the pipelines and the depth of the trenches required. Only with this information could a precise cost estimate be provided. As the commission returned to Europe in March 1892, the government assigned Barois to oversee the project, a predictable outcome given how he had shaped the commission's recommendations.

The Detailed Study

Born in 1849, Julien Hippolyte Eugène Barois graduated from the prestigious French engineering school École Polytechnique in 1868. In 1880, he participated in the imperial mission to build the Trans-Saharan Railway, intending to connect Algeria with West Africa—the French response to the British Cape-to-Cairo railroad project. The following year, Barois moved to Egypt and founded the Société anonyme égyptienne d'entreprises et de travaux publics (Egyptian Enterprise and Public Works Company Limited). In 1884, he was appointed General Secretary of the Egyptian Public Works Ministry, a role he held until 1906. After returning to France in 1909, Barois became a member of the Académie des sciences coloniales (The Academy of Colonial Sciences). He authored publications on a range of topics, including irrigation, Cairo's climate, and petroleum prospecting in the Red Sea.[27]

In April 1892, the Public Works Ministry created a special bureau to conduct the detailed study with Barois as director.[28] The design bureau included engineers like Jean-Baptiste Chiarisoli, director of the Tanzim Department, which managed a range of urban planning and engineering issues. It also included cartographers like H. Ravon, Head of the Map Drawing Office of the Public Works Ministry, whose office was responsible for producing officials maps before the creation of the Survey Department in 1898. Another member of the bureau was the Egyptian engineer Mahmud Fahmi, whose project received an honorable mention from the international commission. Fahmi would later become a building inspector at the Tanzim Department, a member of the Khedivial Cabinet, and the director of works at the royal palace of Hilmiyya.[29]

The design bureau's most pressing task was to produce the detailed map. In 1892, Cairo's authorities had only one map suitable for general planning and engineering purposes—the one made in 1874 by Grand Bey. Nearly two decades old, this map had become outdated. For instance, it partially represented Bulaq and Old Cairo and entirely omitted 'Abbasiyya—areas where the city was currently expanding.[30] Additionally, the 1874 map was drawn on the scale of 1:4,000, which, although adequate for general purposes, fell short in terms of infrastructural design. Barois and the bureau needed a much higher level of precision to accurately determine the placement of sewer pipelines.

As indicated above, the economic policies of the colonial regime began to shift in 1892 with the admission that Britain was to stay in Egypt. As the British administration began considering massive long-term projects such as the Aswan Dam, it sought to reorganize the bureaucracy accordingly. One of the most significant changes concerned the country's taxation system, which colonial administrators generally viewed as arbitrary and inefficient.[31] Cairo's existing tax registers were composed of written records based on *hujjas*—documents that verbally described the location of landed property and recorded transactions, ownership, or occupancy. The Cairo drainage project provided an ideal opportunity to modernize these registers with the use of maps, replacing words with exact spatial information. Specifically, the Public Works Ministry asked Barois and the design bureau to produce a map that could also be used as a cadastral survey. The goal was to draw each plot of land in Cairo and list the name of the property owner on the map itself, thereby creating a cartographic register of land titles. The new register would facilitate the Finance Ministry's mission to raise the house tax from 8.3 to 10 percent of annual rental value, and to direct the added revenue toward the city's drainage scheme.

Barois and the design bureau needed to produce the map as quickly as possible, as it was necessary both for design purposes and to reach an accurate cost estimate. As a result, the map was made in a hurry during the second half of 1892. While the design bureau surveyed parts of the city, it also relied on existing drawings used by the Tanzim Department—plans that were typically used for road construction, repair, and realignment. The bureau gathered these plans, standardized their scales, and filled in missing information. Hence, as historian Jean-Luc Arnaud noted, the Barois map was more like a collage of drawings than something made from scratch.[32] This hybrid approach gave the Barois map its unique characteristics.

When the map was finished, it stood out as an unprecedented cartographic undertaking. It was remarkable in both size and scale, consisting of 353 sheets: 297 sheets at a scale of 1:200 (each centimeter on the map represented two meters of physical space), 53 sheets at 1:500, and

3 sheets at 1:1,000.[33] The 1:200 scale was reserved for the densely built old neighborhoods (see Figure 11 above). The 1:500 scale was applied to the sparsely populated areas developed since 1867 (Figure 14). The map identified every urban feature using names and colors. Communal water facilities—mosques, public baths, and fountains—were outlined in green, while wells and standpipes were marked in blue.

One of the most striking aspects of the map is the irregular contours and orientation of the sheets. At first glance, this seemingly chaotic arrangement can be disorienting. In fact, it was a result of how the map was based on current administrative borders. Predating British rule, the *shiyakha* was the smallest administrative unit in Egyptian cities, typically consisting of a few adjacent streets. Because each sheet corresponded to a *shiyakha* or combinations thereof, the orientation of the sheets was largely dictated by the shape of the streets as well as other prominent topographic features.

The map recorded landownership not only of individuals but also of religious foundations and the state. In some cases, irregularities were noted,

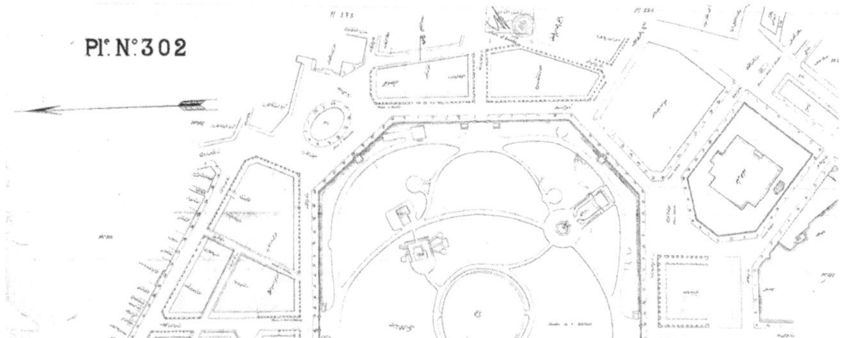

FIGURE 14. The 1892 Cairo survey, sheet no. 302 (Azbakiyya). The Egyptian Survey Authority.

such as land encroachments by slums, or instances where ownership was unknown or disputed. In other cases, they simply listed the owner as *ahali* (people) (Figure 15). By the end of 1892, the Egyptian government possessed an updated database of land titles in Cairo, complete with names of landowners and the limits of their properties shown on the map.

Once the map was finished, the design bureau shifted focus to another critical task: the land-level survey. By installing hundreds of benchmarks throughout the streets, the office collected extensive data on the elevations and depressions across all public roads.[34] This information was summarized as contour lines on a reduced version of the 1892

FIGURE 15. The 1892 Cairo survey, extract from sheet no. 282, depicting an area south of Sayyida Zaynab that was recently developed after draining a pond. The Egyptian Survey Authority.

map (Figure 16).[35] In effect, the bureau successfully created a detailed, three-dimensional representation of Cairo's surface. This data was essential for determining the precise placement of the sewers, a crucial factor in the success of the gravity-based approach. By laying sewer pipeline at the right angle, the network would have enough gradient to maintain efficient flow without the need for motorized power.

The design bureau conducted further research and experiments in preparation. First, Barois gathered information on existing water and gas pipelines, which was important to avoid damaging existing infrastructure.[36] Second, he proposed opening several new streets to facilitate sewer construction, including one in place of the Khalij Canal and another between Azbakiyya and Bab al-Futuh, cutting through dense urban fabric.[37] Third, the bureau conducted various experiments on building materials, concluding that brick sewers should not be used due to the porous nature of Cairo's soil.[38] Fourth, Barois planned to conduct trials on connecting homes in the old quarters to the sewer system, but the idea was dropped due to a lack of funds.[39] Lastly, Barois examined potential locations for the sewage farm, settling for a site in Matariyya. But as will be shown next, the location was later changed due a successful campaign by local residents and neighboring villages.[40]

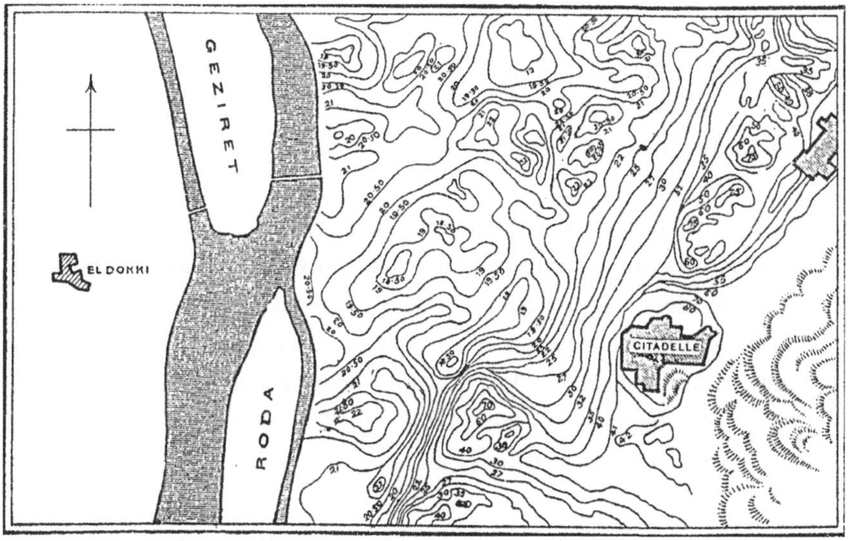

FIGURE 16. Contour map of Cairo on the scale of 1:4000. Clerget, *Le Caire*, Vol. 1, 46.

Not in My Backyard

Matariyya was then a remote village eight kilometers northeast of Cairo, situated on the ruins of the ancient city of Heliopolis, where the cult of the sun god Atum was originally established. In the 1880s, the village was connected to Cairo via road and rail. By 1894, it had evolved into a popular suburb known for its fresh air, leading several doctors to recommend it as a health resort.[41] At the same time, Matariyya became a tourist destination, boasting attractions such as the oldest standing obelisk in Egypt (erected by Senusret I), the Virgin's Tree, a petrified forest, and an ostrich farm. The suburb attracted a mix of wealthy Europeans and Egyptians who appreciated the clean air and semi-rural environment.

In early 1894, residents of Matariyya learned that their cherished neighborhood had been selected for Cairo's planned sewage system. The Public Works Ministry intended to convert thirty square kilometers of land bordering the suburb into a sewage farm. The announcement raised concern among the residents, not only because of the association between disease and miasma, but also due to the farm's potentially negative impact on the value of their property. Accordingly, a group of residents organized a campaign to oppose the plan. They first submitted a collective petition to Khedive 'Abbas Hilmi—the great-great-grandson of Mehmed 'Ali and the last ruler of Ottoman Egypt—appealing for his intervention to halt the project. The petition began with a brief explanation of why they had moved to the suburb in the first place. Health concerns, they revealed, were of utmost importance:

> As Cairo's population grew, health conditions deteriorated, with deadly infectious diseases spreading more each year. In response, many residents sought a safe refuge to protect themselves and their children from the ravages of disease. Some chose Hilwan, while we went to Matariyya. . . . We invested heavily and endured the challenges of relocating to such a barren place to enjoy clean air and safety from the city's sickness and decay.[42]

The residents argued that the chosen location of the sewage farm would be detrimental to the health of not just Matariyya and surrounding

Ultimately, the cadastral survey of 1892–1907 enabled the British regime to abolish the old tax classifications of *kharaj* and *'ushr*, replacing them with a unified tax rate based on the new valuation.[62] This is particularly noteworthy in light of the significant increase in land values following the construction of the Aswan Dam. In this way, the British regime was positioned to extract a portion of the surplus generated by its own development projects and economic policies.

Despite the obvious similarity between the Cairo survey of 1892 and the national survey of 1892–1907, there is a notable difference in their organizational principles. As indicated earlier, the Cairo survey was created in a rush and under pressure to carry on with the Cairo drainage project. The decision to divide the maps along the physical border of *shiyakhas*—the smallest administrative urban unit in Egypt—should be understood in this context. However, when combined with the inconsistent scales and orientations, this approach made it challenging to assemble even few sheets together. Therefore, the map proved too cumbersome to use for any other purposes.[63]

On the other hand, the national survey of 1892–1907 was organized on a different basis. Instead of relying on physical features and administrative borders, the Survey Department began from an abstract geometric grid that covered the entire country, dividing national space into equal squares that were further subdivided into smaller equal squares.[64] In this way, the Department created a comprehensive cartographic system for the whole country, enabling the easy combination, division, and rescaling of maps as needed. Thus, while the 1892 Cairo survey was an important precedent for the 1892–1907 national survey, it also served as a cautionary tale, highlighting the need for a more systematic approach capable of accommodating a wide range of cartographic needs. Whereas the former was geared toward a specific engineering project, the latter was an expression of the British regime's desire to standardize national space as a whole.

Cairo's Geology

In fairness to Barois and the international commission, the project's failure was not only a matter of oversight but also of limited geological knowledge. In 1892, the composition of Cairo's soil up to about six meters below the surface was familiar to scientists and ordinary people alike. Below that depth, however, knowledge rested largely on conjecture and speculation. Had such knowledge been available, Barois and the commission might have reached a more precise cost estimate from the beginning. This situation began to change in 1895, when the British administration conducted a comprehensive geological survey. Prior to the survey, geological knowledge advanced largely on the margins of commercial, military, and archaeological interests. With the survey, the government aimed to take charge of geological research and direct it toward the extraction of Egypt's mineral resources. In this sense, it was part of the British colonial regime's broader modernization and state-building efforts. At the same time, the survey helped uncover the deep geological history of the Nile Valley.

Geological knowledge predating British rule was already quite sophisticated. Cairo's very location and its traditional architecture reflect an advanced understanding of the soil. The city's Fatimid nucleus was built on the elevated edge of the desert, safely beyond the reach of the Nile's annual flooding. The location struck a balance between the dryness of the soil and the ability to extract water from wells. With every flood season, the Nile rose by six meters on average, causing water to infiltrate the city's porous soil. The rise and fall of water in wells provided a common-sense understanding of the water table and its connection to the Nile. In low-lying areas, the proximity of subsoil water was well known, sometimes even a visible fact, which is why these areas remained mostly vacant until they were developed after 1867. Traditional houses in Cairo were designed with an understanding of these environmental conditions in mind. Ground floors were constructed with thick limestone to resist moisture, while lighter bricks were used for the upper floors. Percolating cesspools also relied on the absorptive properties of the soil and the annual "flushing" caused by the rise and fall of the water table. At the same time, the construction of large mosques and multi-story caravanserais required close familiarity with the nature of the soil.

Throughout the nineteenth century, several deep borings that combined scientific and practical interests were carried out in Egypt. Under the rule of Mehmed 'Ali Pasha, the French engineer Linant de Bellefonds drilled deep holes in preparation for building the Delta barrage, one of the most ambitious irrigation projects of the time.[65] In the 1830s and 1840s, additional borings were made to prospect for underground water at greater depths, though the water obtained was similar to that from shallow wells.[66] Other borings encountered thick layers of clay that could not be penetrated.[67]

Between 1851 and 1854, Scottish geologist Leonard Horner collaborated with Armenian-Egyptian engineer Joseph Hekekyan to drill deep holes at the sites of the ancient cities of Heliopolis and Memphis.[68] Their goal was to apply stratigraphy—the study of the sedimentary layers of the earth—to archaeology. By analyzing the soil at great depths, Horner hoped to estimate the antiquity of Egyptian civilization and identify the boundary between geological time and human history.[69] In the 1840s, Italian geologist Antonio Figari led a series of mining expeditions in search of marble and coal. Figari is credited with creating the first geological map of Egypt, although it was later found to be inadequate.[70] Finally, under Khedive Isma'il's urban development program, the construction of massive buildings like the Railway Station and the Opera House required deeper excavations than usual.[71] These activities revealed that the topsoil beneath Cairo was a mix of sand, silt, rubble, bones, plant debris, potsherds, and remnants of ancient canals, cemeteries, and battle sites.[72]

Other geological studies were conducted during the first decade of British rule. Between 1883 and 1886, the Royal Society of London established a special committee to study the geology of the Nile Delta. This "Delta Committee," composed of geologists and engineers from the British army, carried out deep borings in Kafr al-Zayyat, Tanta, and Zaqaziq. Their goal was to drill down to what they assumed to be the rocky floor of the Nile Valley, but they found only sand, silt, and clay at depths ranging from fourteen to twenty-six meters.[73] As the sewage system was being considered in 1890–92, the Public Health Department collected data from three different wells in Cairo and used it to create a diagram illustrating fluctuations in subsoil water levels. In 1894, the Cairo Tanzim Department began weekly observations of the water table in six special wells.[74] This new set of data largely confirmed the connection between the Nile and the rise and fall of the water table.

In 1895, the government conducted a geological survey of Egypt, directed by Henry Lyons, who later became head of the Survey

Department. The survey aimed to gather reconnaissance on all mineral resources of economic value, including water, oil, salt, nitrates, and phosphates. The government had recently received several business proposals for oil and mineral prospecting, but it lacked the necessary information to evaluate them. As Lyons explained in a memorandum, the government was "working in the dark."[75] Only by locating and describing the country's mineral resources could they be made commercially available. In effect, the government facilitated and coordinated the extraction of these resources for the benefit of private business.

Meanwhile, the survey also shed light on the deep geological history of the Nile Valley. One of its early conclusions was that the valley dates back to the late Miocene and early Pliocene eras—roughly six million years ago—when a catastrophic event caused extensive fracturing of the earth's crust. Lyons described how, during this period, the sea

> laid down thick deposits of sand and gravel on the floor of the valley and up to the foot of the cliffs bounding it, while the tributary streams, fed by rainfall much heavier than that of to-day, brought down masses of detritus from the limestone plateau and piled them up along the margins of the valley. A subsequent rise of the area converted this "fiord" into a river valley, and the deposition of the Nile mud and the formation of cultivable land began.[76]

In other words, the soil of the Nile Valley consists of layers of gravel, sand, silt, and clay, deposited downstream by annual floods over the course of many thousands of years. Lyons estimated the rate of alluvial accumulation at approximately twelve centimeters per century.[77]

By the end of the nineteenth century, geological studies in Egypt had provided a clearer understanding of the soil structure of the Nile Valley. The first six meters typically contained variable mixtures of sand and silt. Below this lay a second layer, whose thickness and composition varied greatly; in some areas, sand was predominant, while in others, clay was the main component. At greater depths, a third layer of compact clay was found in many areas on both sides of the valley. Efforts to penetrate this clay stratum had so far been met with considerable resistance.

Egyptian Engineers Under British Occupation

Egyptian engineers found themselves relegated to minor roles or entirely sidelined from large projects that were studied and implemented by the British colonial administration. The extensive repairs to the Delta Barrage were overseen by Scott-Moncrieff. The Aswan dam project was designed by the British irrigation engineer William Willcocks. Julien Barois was put in charge of the Cairo drainage project with Mahmud Fahmi serving under him as the only Egyptian in the design bureau. These and other cases demonstrate the systematic marginalization of indigenous professionals under British rule. Excluded from government decision-making, Egyptian engineers began to organize outside its purview. A significant landmark in this effort was the establishment of *al-Muhandis* (The Engineer), Egypt's first Arabic-language engineering journal.

Founded in 1893 by Ahmad Kamil, an Egyptian engineer in the charitable endowments administration, the monthly journal *al-Muhandis* emerged in a unique context. Earlier that year, Willcocks stirred controversy with a speech at the Azbakiyya Club in Cairo. Known for his prominent role in Britain's irrigation program in Egypt, Willcocks was also a keen advocate of the colloquial Egyptian dialect.[78] When not designing irrigation works, Willcocks found the time to translate the New Testament into colloquial Arabic (rendered as *al-Khabar al-Tayyib bita' Yasu' al-Masih*) and to commission similar translations of European classics. He even authored books in Egyptian vernacular, such as *Al-Akl wa al-Iman* (*Food and Faith*). In his Azbakiyya Club speech, he claimed that Egyptians lacked the "power of invention" because they wrote in a "dead language" (classical Arabic), which, he argued, was no longer effective for transmitting knowledge beyond a small, formally-educated audience.[79] The speech was later published in *al-Azhar*, a journal he co-edited with Ahmad al-Azhari, an Egyptian civil servant.

Given the deep cultural connection between classical Arabic and the holy text of Quran, many Egyptian intellectuals quickly went on the defensive. 'Abdullah al-Nadim, an influential writer who played a leading role in the 'Urbai revolt of 1882, ridiculed the notion of abandoning classical Arabic.[80] Launched shortly after the speech, *al-Muhandis* explicitly stated its commitment to preserving the classical language "despite

Willcocks's efforts."[81] Declaring its mission to spread "the science of engineering in the honorable Arabic language," the journal covered a broad range of topics, including construction material, architecture, railways, steam engines, nature, and astronomy.[82] To Willcocks's credit, his journal *al-Azhar* published several responses that criticized the lecture. Among these was a response authored by Ahmad Sulayman, an engineer at the Tanzim Department. While thanking Willcocks for his "noble passions," Sulayman argued that the lecture was misguided because the Egyptian dialect was not a separate language but rather a deviation from classical Arabic. More convincingly, Sulayman pointed out that it was misleading to discuss the "power of invention" in the abstract, without addressing the educational institutions required to support it. He maintained that it was, in fact, the British colonial government that stifled the power of invention by underfunding the Egyptian engineering school (*al-Muhandiskhana*) and technical education in Arabic more broadly.[83] On this account, Sulayman was undeniably on point.

Willcocks was generally respected by many Egyptians, perceived as more sincere than other British officials, or at least more palatable. After facing significant public criticism, he discontinued the journal *al-Azhar*. He was also reprimanded by Lord Cromer for publicly voicing his opinions on sensitive matters in the local press.[84] In hindsight, the outcome of this exchange was largely predetermined by Willcocks's position as a British engineer in occupied Egypt. The debate over diglossia is still a contentious issue in the Arabic-speaking world. Today, many voices advocate for the study and codification of vernaculars, without necessarily undermining the classical language.

Cholera Strikes Again

The Cairo drainage project was officially shelved by the end of 1894. Only one year later the Egyptian government was reminded yet again of the project's vital nature from the standpoint of public health. A cholera epidemic broke out in 1895–96, spreading across Cairo, Alexandria, and parts of the Delta. Compared to the 1883 epidemic, the British administration was better prepared and more convinced of the need to follow the latest scientific advice. By this time, Robert Koch's discovery of the role of contaminated water in transmitting cholera had been repeatedly confirmed, most recently during an outbreak in Hamburg, Germany.[85] From that point onward, no city could afford to neglect the scientific regulation of its water supply without risking public health. At the same time, Koch also emphasized that improved sanitation, including water filtration and proper waste disposal, was crucial to preventing outbreaks.

The government's response to the 1895–96 epidemic was markedly different from that of 1883, according to Rogers Pasha, director of the Public Health Department.[86] Protecting the water supply had become central to the government's policy. Several measures were implemented to this end. The government barricaded sections of the Nile banks in Bulaq and Old Cairo to prevent unauthorized water drawing and installed standpipes to freely distribute filtered water. Sailboats were prohibited from approaching the water intake sites, and water carriers were prevented from drawing water without government authorization. Mosques were instructed to change the water in ablution courts more frequently, with some being shut down for failing to meet sanitary standards. The government also destroyed water wells deemed unsanitary or located too close to cesspools. Three makeshift hospitals were established in Old Cairo, Bulaq, and 'Abbasiyya, where patients and their families were isolated and monitored for six days before release. As a result of these measures, the number of fatalities in Cairo was 1,343, significantly lower than in 1883.

The epidemic led to a violent standoff between the police and students at al-Azhar University, one the most distinguished seminaries in the Islamic world. In June 1896, health authorities discovered that a Syrian student at al-Azhar had contracted cholera. Government doctors attempted to inspect the student and transfer him to an isolation hospital, but his colleagues refused to hand him over, telling the university president that "all who go to the hospital die."[87] Tensions escalated when government officials called in the police and the students reacted by throwing rocks, striking the Cairo governor and the British Deputy Chief of Police. In response, the police opened fire on protestors, killing three and wounding two others (Figure 17). Hundreds of students were subsequently arrested, and the government shut down the Syrian hall at the university for one year. Additionally, the president of al-Azhar University was reprimanded for failing to use his influence to quell the unrest.[88]

Ce numéro est accompagné d'une gravure hors texte en couleurs.

L'ILLUSTRATION

Prix du Numéro : 75 cent. SAMEDI 27 JUIN 1896 *54e Année. — No 2783*

LE CHOLÉRA AU CAIRE. — Révolte d'El-Azhar : la police égyptienne faisant feu sur les étudiants.

Dessin d'après nature de M. Philippoteaux.

FIGURE 17. Cholera riots at al-Azhar. *L'Illustration*, June 27, 1896.

Much like in 1883, the government's anti-cholera measures in 1895–96 were deeply unpopular and surrounded by rumors and conspiracy theories. The outbreak presented an opportunity for the government to finally fill in the Khalij Canal—a plan long considered necessary for public health but repeatedly postponed due to the canal's popularity. With widespread recognition of the link between water and disease transmission, the government judged it was time to implement this plan. In September 1896, the government dammed the Khalij, prohibited water collection, and disinfected the entire canal with lime. By the end of the month, when the Khalij would typically be filled with water, its bed was completely dry.[89] Two years later, the canal was transformed into a street with a tramline.

Rogers recounted that locals avoided the canal during the disinfection process, believing that "the water was being poisoned by the Sanitary Department."[90] For him, the rumor worked to the government's advantage, as it discouraged people from using the canal. Another rumor suggested that sanitary officials sedated people by throwing handkerchiefs over their faces, then abducted them to hospitals from which they never returned.[91] As Africanist historian Luise White argued, such rumors gave voice to the experience of exploitation under colonial rule.[92] While clearly false, they nevertheless reveal the vulnerability felt by the colonized as they were being subjected to invasive forms of medical intervention without sufficient explanation or consultation.

Conclusion

By the 1890s, British colonial administrators were well aware of the need for drainage in both cities and the countryside. With the government's irrigation program in full force, water was being applied to agricultural land in greater volumes. However, without adequate drainage, the soil became saturated, causing salt to rise to the surface and negatively affecting crop yields. Alarmed by these developments, the government embarked on serious efforts to dig agricultural drains. In his 1893 annual report, Cromer highlighted the significance of this work:

> It is impossible to over-estimate the benefits, direct and indirect, caused by the better drainage of the country. The results are seen in

> the way in which the taxes now come in, and in the increased yield of the crops in the Deltaic provinces. . . . Much has certainly been done, but very much yet remains to do; the ordinary Budget is only sufficient to meet the needs of irrigation works; drainage works are costly, owing to the heavy cubes of earthwork, and to the massiveness of the masonry works required for carrying them under the network of existing canals.[93]

While urban drainage was driven primarily by public health concerns, its scale and cost were no less daunting. As we saw, Barois learned this the hard way. Visualizing sewer pipelines on a map was one thing, laying them at great depths beneath Cairo's surface was another. Cutting through layers of stiff clay and stabilizing pipes in saturated soil proved to be a far greater challenge than anticipated. Consequently, the Egyptian government ended up with a plan that was twice as expensive as the one rejected a few years earlier.

This chapter has traced the history of the Cairo survey of 1892, showing how it was intended to help design and fund the city's drainage project. It also argues that the survey was part of a wider effort by the British colonial regime to develop the country's resources and upgrade its bureaucratic apparatus. Another element of this program was the geological survey of 1895, which was significant for both scientific and economic reasons. That survey aimed to address gaps in knowledge about the composition of Egyptian soil while facilitating the commercial exploitation of mineral resources such as water, phosphate, manganese, and oil.

The resurgence of cholera in 1895–96 provided a chance to apply the results of the geological survey to areas where it was critically needed. The epidemic intensified anxieties over the purity of Cairo's water supply, placing mounting pressure on the colonial government to respond. While barricading the Nile banks may have been effective during the crisis, it was unsustainable in the long run. The outbreak underscored the troubling reality that the Nile was always liable to contamination. As the following chapter shows, these circumstances led the government to encourage water supply companies to develop the country's underground water resources. The success of this initiative depended on finding water deep enough to be free from surface contamination.

FOUR

Water and Land

Egypt's Hidden Nile and the Building Boom in Cairo

Badi' Khayri, a folk lyricist and playwright, is known in Egyptian popular culture for his collaboration with Sayyid Darwish, a pioneering musical figure, on a series of satirical songs that voiced the grievances of craftsmen and service workers in Egypt.[1] One of these was "The Water Carriers' Melody," written in 1918:

> May God ease the burden of the water carriers and compensate them; they are exhausted and frustrated with the company [the Cairo Water Company]
> The company's foreigners have come to drive us out. Why are they harassing us when this is our ancestral trade? Don't ignore us, people.
> This company is a nuisance; its water is filthy. It's murky, green, and blue, though they claim it's pure
> The company's water is disgusting; they add carbonates, wine, sulfur phosphides, and even genie's powder; haha.[2]

Inspired by folk stories and popular street songs, Khayri captured the water carriers' bitterness toward the Cairo Water Company, which had been eroding their profession for decades. In the song's third line, the carriers made a curious claim about the color of the company's water, using the word *nigsa* (derived from *najis* in classical Arabic, meaning ritually impure), suggesting that the water was unfit for religious ablution—a barb harkening back to the time when faucets were first installed inside mosques (Chapter 1). The fourth line went even further, alleging that the company added chemicals and magical powders to the water.

This playful exaggeration evoked the recent debacle of 1905–1910, when the water company, in consultation with the government, abandoned the Nile as the source of the city's water supply and began distributing water drawn exclusively from deep wells. Promoted as a groundbreaking application of science and technology to public health, the project failed spectacularly after a five-year trial, largely because of widespread complaints about the taste and perceived purity of the new supply. The criticisms were varied and often puzzling: some claimed the well water was yellow, brown, or even purple, and difficult to digest. Women complained that it damaged their hair, yellowed linens, and was unfit for lathering soap or cooking beans. Described as "dead and buried," well water was even rumored to cause sterility. Under mounting public pressure, the new supply was discontinued, and the company returned to the water of the Nile.

The decision to change the water source had been in the works since 1897, when the Société Anonyme des Eaux de Tantah (Tanta Water Company) successfully extracted what appeared to be pure water from deep wells for the first time. Government officials and investors hailed this achievement as marking the dawn of a new era for the water supply industry. Unlike the Nile, which was always susceptible to contamination, deep well water was believed to be safely beyond the reach of germs. At the same time, the discovery fueled speculation—often exaggerated—about the extent of groundwater beneath the Nile Valley. Some believed that there was a vast reservoir of pure water underneath Egyptian soil—a "hidden Nile" waiting to be developed and exploited.

This chapter places these events within the broader context of an unprecedented economic bubble in Egypt, spanning the decade between

1897 and 1907. Excessive enthusiasm about water resources was not unique in itself but rather a symptom of a larger speculative euphoria that swept across the whole country. Roger Owen was one of the first to study the impact of this decade-long cycle of boom and bust on the Egyptian economy, showing how it exposed the risks of cash-crop agriculture and led to new ideas of economic diversification, industrialization, and calls for greater government intervention, including the establishment of national banking institutions.[3] Owen also examined the boom's specific effects on Cairo, revealing how it triggered a construction frenzy, the extension of tram and railway lines, suburban sprawl, and a sharp rise in real estate values.[4] More recently, Aaron Jakes has offered a new perspective on the bubble, tracing its origins to the economic policies of the British colonial regime in Egypt.[5] As he persuasively argues, the bubble was an unintended consequence of British plans for economic development. Colonial administrators imagined that improved irrigation, simplified taxation, mortgage banking, and expanded cotton production would foster a society of satisfied small landholders. The bubble, however, unleashed dynamics that instead deepened peasant poverty and indebtedness.

Building on these contributions, I argue that Cairo's transformation during this decade should be analyzed not only as a spatial phenomenon but also in hydrological terms. By integrating land and water into the same analytical framework, this chapter brings into view a key development that has been largely overlooked by the city's historians.[6] The boom produced a new generation of buildings that differed significantly from the old, not only with respect to architectural aesthetics but also in terms of indoor plumbing and connectivity to the water supply. Whereas traditional houses were not designed for heavy water usage, those built after 1897 were increasingly designed with domestic water consumption in mind, featuring novel appliances such as flushing toilets, kitchen sinks, bathtubs, drains, and waterproof cesspools. In essence, the home was reconfigured to accommodate greater levels of private water usage. With the boom in full swing, entire new suburbs were planned with integrated water and sanitation systems. Compared to the older two-story houses, the new six-story apartment buildings required more water and higher pressure. By 1907, approximately 15 percent of buildings were connected to water mains, up from 7.5 percent in 1892.[7]

These changes were widely understood to reflect "modern" norms of bodily hygiene and privacy and were often framed in civilizational terms. Commenting on this transformation, French Egyptologist Georges Foucart claimed that Cairo's housing conditions were being brought "up to date" to align with those of European cities.[8] Yacoub Artin, an Armenian-Egyptian scholar and former minister of education, described it as a textbook illustration of the saying that "the luxury of today is the necessity of tomorrow."[9] Alfred Eid, a Syrian-Belgian doctor and investor, argued that the demand for water, sanitation, and privacy arose from "more intimate contact with Europe."[10] The idea that Egypt was now beginning to catch up with European modernity, or at least with its trappings, was common sense in elite circles. In fact, however, these developments emerged from a more intimate encounter with death, one shared across Europe and Egypt alike. As repeated cholera pandemics made clear, water and sewers were no longer luxuries, but essential infrastructures of public health.

Leaders of the construction and housing industry claimed that they were doing service to the public by expanding the supply of healthy housing equipped with improved water and sanitary facilities. However, Egyptian voices like the young journalist Ahmad Hilmi argued that the industry was driven primarily by profits rather than genuine concern for people's health. Hilmi pointed out that there was a significant gap between what the developers offered and what most residents actually needed to improve their lives. He criticized the construction boom as a form of "mobile urbanization," in which investors rushed from one area to another in search of higher returns.[11] While earlier chapters have focused on the roles of government, the water company, and disease in driving urban change, this chapter shows that the housing industry was an equally consequential actor.

The Hidden Nile

By the nineteenth century's end, Egypt had entered the era of deep drilling and water prospecting. So far, efforts to tap deeper into groundwater had been repeatedly hindered by a thick layer of compact clay that seemed to lie beneath many parts of the Nile Valley. However, in 1897, German engineer Karl Abel, working for the Tanta Water Company, succeeded in drilling through the stiff clay and drawing water from underneath. This breakthrough set the stage for a decade of speculative optimism, with officials, engineers, and water experts proclaiming the existence of an inexhaustible reservoir of water hidden under the Nile Valley.

The Tanta Water Company was an early beneficiary of the government's efforts to promote the economic exploitation of Egypt's mineral resources (Chapter 3). Founded in 1895, the same year the government initiated the geological survey, the company secured a fifty-year concession to supply water to the country's third-largest city, after Cairo and Alexandria. The venture was financed by prominent members of Egypt's leading business elite: Felix Suares, Jacques Menasce, and Yusuf Qattawi.[12] The three came from Jewish families that had initially established themselves as moneychangers and lenders in Muski, Cairo's Jewish quarter. Collectively and individually, they held diverse investments in banking, real estate, railways, and sugar production. In 1880, the Suares family partnered with French investors to create the Crédit Foncier Égyptien (Egyptian Mortgage Bank), which became the country's leading agricultural credit institution. They also owned the Cairo Omnibus Company and significant land in the upscale Isma'iliyya district.[13] During the 1890s, the three families expanded their investments into the water supply of provincial towns, including Tanta, Zaqaziq, and Mansura.

The Tanta Water Company was founded on a completely different premise than those in Cairo or Alexandria. Instead of relying on the Nile or canals, the Tanta Water Company planned from the outset to supply water drawn from underground sources. Following the cholera epidemic of 1895–96, government officials began considering the idea of abandoning the Nile as a water source. Around this time, William Garstin, advisor to the Public Works Ministry, approached financiers, including Suares, who were willing to fund water prospecting efforts.[14]

The Tanta Water Company hired German engineer Karl Abel to conduct the necessary experiments. Abel had spent several years working in Italy, specializing in well sounding—the measurement of water depth in wells. In Egypt, he applied a method of earth sensing that was being tested in other countries, including the United States and Algeria.[15] Assisted by an unidentified "amateur geologist," Abel carried out four experimental borings around Tanta.[16] He employed a double-piping system: first, a large pipe was forcefully driven into the ground, the soil inside removed, and then a smaller pipe, equipped with a sieve at the end, was inserted and lowered into the ground (Figure 18). The larger pipe cut through the earth and collected samples of geological strata, while the smaller one collected water, if present. Until then, attempts to drill deep wells had either been thwarted by a thick layer of clay or failed to access water distinct from shallow infiltration sources.

The first experimental boring was conducted between July and August 1895, near the Ja'fariyya Canal, which had long been the town's primary water source. Abel drilled the pipe to a depth of twenty-eight meters, cutting through layers of silt, sand, and dense clay. As soon as the clay was pierced, water began streaming out of the pipe. The second boring, carried out in October of the same year, produced similar results. The third boring, in November, reached a depth of thirty-one meters, with water gushing out at a rate of one liter per second. The fourth boring, completed between January and February 1896, extended forty meters below Tanta.[17] Observers believed that Abel had successfully dug the first "artesian" wells in Egypt. An artesian well is a specific type of aquifer in which water is trapped beneath layers of earth under positive pressure. When the aquifer wall is punctured, the pressure forces water out naturally, without the need for pumping, unlike traditional tube wells.

This discovery marked the beginning of a new era, declared Yacoub Artin in an address to the Institut Égyptien, a learned society founded in 1798 during the French Expedition in Egypt.[18] Artin, a public servant and scholar with diverse interests ranging from folk tales to the latest technological and economic advancements, was a firm believer in developing Egypt's underground water resources. Earlier that year, he visited Tanta's wells and discussed the findings with Abel. In his speech, Artin argued that the use of artesian wells should be extended to other Egyptian cities, provided similar geological conditions existed. He noted that the discovery had already attracted the attention of Alexandria's water company and expressed hope that the Cairo Water Company would also consider the possibility.

In Artin's view, what happened in Tanta indicated the existence of a vast reserve of pure water beneath Egypt's soil. In a subsequent address,

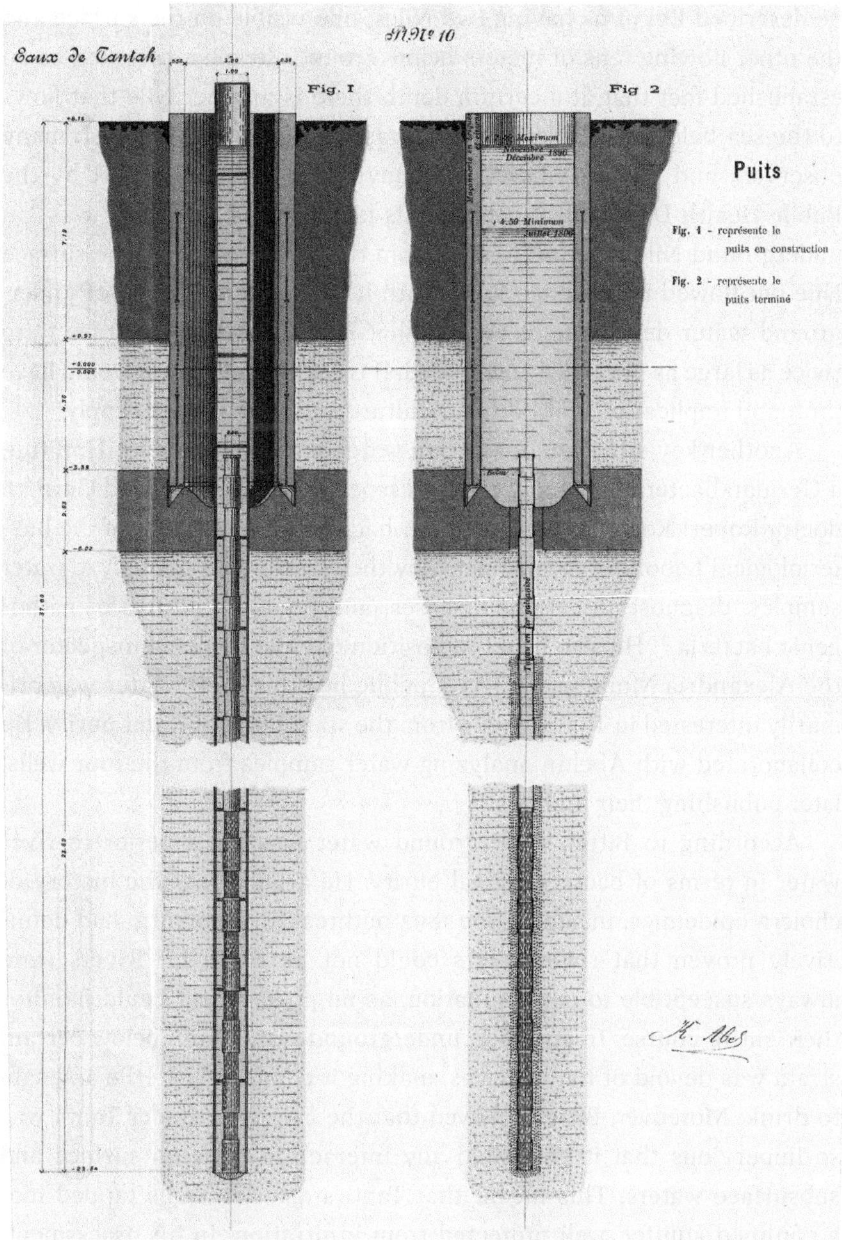

FIGURE 18. The double-piping system used by Karl Abel in Tanta. Bitter, *Observations hygiéniques*, plate no. 10.

he described Egypt as having two Niles, one visible on the surface and the other flowing tens of meters below ground: "It must be taken as an established fact that at a certain depth there is another Nile that flows to the sea below the valley."[19] Artin was not alone in this belief; many observers and officials shared his conviction.[20] A 1906 report by the Public Health Department echoed this idea, stating that there was "an underground Nile" that originated from the same source as the surface Nile but flowed independently beneath it.[21] Thus, proponents of underground water development argued that Egypt's water resources were twice as large as previously estimated. If true, this revelation would have profound implications for both agriculture and urban water supply.

Another key advocate of artesian water supply was doctor H. Bitter, a German bacteriologist and former associate of the celebrated German doctor Robert Koch. Since 1894, Bitter had served as director of the Bacteriological Laboratory, established by the government to analyze water samples, diagnose infectious diseases, and prepare cultures of pathogenic bacteria.[22] He also held the position of chief hygienic inspector of the Alexandria Municipality. As a public health expert, Bitter was primarily interested in Tanta's wells from the standpoint of water purity. He collaborated with Abel in analyzing water samples from the four wells, later publishing their findings.[23]

According to Bitter, underground water was far superior to river water in terms of bacteriological purity. He argued that the history of cholera epidemics, including the 1892 outbreak in Hamburg, had definitively proven that river waters could not be trusted.[24] Rivers were always susceptible to contamination, as no government could monitor their entire course. In contrast, underground water from below certain strata was devoid of living germs, making it completely sterile and safe to drink. Moreover, Bitter believed that the clay layer under Tanta was so impervious that it prevented any interaction between surface and subsurface waters. This meant that Tanta's artesian wells tapped into a confined aquifer, well protected from infiltration. In his assessment, the water from these wells was not only hygienically pure but also clear, odorless, and "very agreeable and refreshing" in taste.[25] Consequently, Bitter recommended the use of artesian wells in Cairo, Alexandria, and other major towns in Egypt.

However, not everyone supported a wholesale shift to groundwater, including British engineer William Willcocks. In 1897, Willcocks became the manager of the Cairo Water Company after years of working for the government's irrigation department. In this capacity, he opposed the use of artesian wells in Cairo, arguing that the idea was inadvisable because the integrity of the clay layer was increasingly being compromised by the building industry: "It had been the practice for some years of the proprietors of the larger hotels and blocks of dwelling-houses to bore through the intervening clay stratum and put the cesspools in communication with the very stratum from which we were to obtain our drinking water."[26] Meanwhile, Willcocks dedicated considerable effort during his two-year tenure as the company manager (1897–99) to drafting a sewage system proposal for Cairo.[27] The proposal was postponed indefinitely due to ongoing negotiations with the foreign Powers regarding the house tax and the Public Health Act, which was still under review by the Mixed Court (Chapter 3).[28] However, it demonstrates the company's strong interest in the sewage system project. Had it been implemented, it would have given the private corporation total control over both ends of Cairo's water cycle at a time when the city was on the brink of significant transformations.

The Building Boom

The notion of a hidden Nile deep under Egyptian soil was just one of many fanciful claims fueling the economic bubble until it burst in 1907. Economic boosters portrayed Egypt as the new El Dorado—a magical place where fortunes could be rapidly amassed.[29] Between 1897 and 1907, the value of foreign capital investments soared from L.E. 13,885,000 to L.E. 87,176,000.[30] Half of this capital was allocated to the land and building industries, while the remainder was invested in mortgage banking, finance, mining, and commerce. Reflecting on the mania, Lord Rathmore, chairman of the Bank of Egypt, remarked that "people were apparently mad; I do not know what other word to use; they seemed to think that every company that came out was worth double its value before it had even started business."[31] This local frenzy was fueled by a global surplus. As Europe became saturated with capital following the 1870s–90s recession, industrialists and financiers sought investment opportunities on the international stage.[32] Egypt was among many countries shaped by the twin forces of empire and international finance.[33]

The bubble was not an isolated event or a historical anomaly. Rather, the panic of 1907 was a global phenomenon that affected financial centers in New York, Santiago, Amsterdam, Cairo, and Tokyo. Because economic manias and crises respect no political boundaries, they have often been likened to contagion.[34] And similarly, the origin, path, and long-term consequences of such sweeping events become subject to intense debates. Nonetheless, there is a broad agreement that economic bubbles share a common pattern.[35] Typically, a new and appealing development, commodity, or innovation—such as railways during the Great Railway Mania of the mid-1840s or the internet during the dot-com bubble of the 1990s—catches the public's interest. Initial gains lead to expectations of future profits, fueling speculative activities in land, shares, and prices. As speculation gains momentum, the bubble expands until it eventually bursts, often resulting in significant long-term economic and social repercussions.

In Cairo, this appealing commodity was housing, particularly that which was designed to meet new standards of hygiene and privacy. During the boom decade, the city's built environment was transformed to accommodate this new mode of living. Canals were filled, tramway lines extended, and poor housing demolished as large suburbs like Ma'adi and Heliopolis emerged from the desert. In the city center, houses were torn down to make way for multi-story apartment buildings, and fashionable neighborhoods like Azbakiyya and Isma'iliyya were rebuilt to maximize space and profit.

Rapid transportation was one of the engines of urban development. Until the mid-1890s, animal-drawn carts and carriages were the predominant means of transit, with the Omnibus Company providing service along fixed routes, and two railway lines connecting Hilwan and Matariyya to the city.[36] The introduction of tramways in 1897 marked the beginning of a new era of mobility and urban growth. The government granted the tramway concession to Baron Édouard Empain, a Belgian investor with a global portfolio of transportation ventures in Belgium, Congo, France, Russia, and China. Within a year of receiving the concession, the Cairo Tramway Company constructed six lines intersecting at al-'Ataba Square—the vital interface between the old city and the new extensions developed since 1867. Over the next few years, additional lines were built, connecting Giza Pyramids, 'Abbasiyya, Shubra, and Rawd al-Faraj.[37] As Egyptian historian Muhammad Sayyid Kilani demonstrated in his pioneering study, the impact of tramways on housing, work, and social interactions was profound.[38]

During the boom era, three of Cairo's most significant waterways were obliterated. The Khalij Canal, long regarded as one of the city's most distinctive yet unsanitary features, was filled in by the Cairo Tramway Company. In 1896, the government drained and disinfected the canal as an anticholera measure (Chapter 3) (Figure 19). A year later, the government granted the company the right to build a tramline on the former canal bed. Under the terms of the charter, the company was required to cover the costs of filling in the canal, construct the road, and compensate homeowners for expropriations. The Public Works Ministry endorsed the project not only for its public health benefits but also for its potential to enhance real estate value: "When a large street carves a straight line into the native neighborhoods, it will increase the value of bordering properties."[39] Similarly, the government filled in parts of the Bulaqiyya and Isma'iliyya canals after a cholera outbreak in 1902, converting them into new streets.

Demolition is the often-overlooked precondition for construction. During the boom, several slums were cleared in the name of public health. But of course, it was also a pretext to make way for development projects geared toward more affluent clients. In previous decades, clusters of poor housing had cropped up in the vacant and less desirable areas between the main urban core and suburbs like Bulaq, Shubra, Old Cairo, and 'Abbasiyya. As land values rose, the government demolished these settlements with startling speed.[40] In *The Egyptian Gazette,* an observer noted that "everywhere one sees new and handsome stone houses replacing the old lath and plaster tenements, and even the mud-built huts of my early recollections."[41]

A significant amount of demolition also took place in the upscale neighborhoods of Isma'iliyya and Azbakiyya, areas formerly known for their spacious palaces and gardens. With the onset of the boom, they

FIGURE 19. The Khalij Canal after it was drained. *The Photographic Memory of Cairo.* Bibliotheca Alexandrina, 2009.

were razed to make room for more intensive development. While the built landscape had previously consisted of two or three-floor structures, the new constructions reached five, six, and seven stories high. Contemporaries noted the rapid pace of transformation: "Whole streets of houses, which were quite new in 1903, were being pulled down [in 1906] in order to make room for new houses with additional storeys."[42] Georges Foucart criticized the new architecture as "brutal and cyclopean," predicting that Azbakiyya would soon resemble a "neighborhood of tall barracks."[43] This vertical growth was facilitated by the increasing use of modern materials, such as reinforced concrete.[44] During this decade, the area developed its characteristic style of multi-story apartment buildings with commercial storefronts. Renowned department stores like Orosdi-Back (later Omar Effendi), Cicurel, and Sednaoui were opened, offering a novel shopping experience to affluent customers.[45]

The most frenzied phase of the boom coincided with the development of several upper-class suburbs, including Garden City, Ma'adi, and Heliopolis. Ma'adi, styled as an "English suburb," was developed by the Suares-led business consortium and Elwin Palmer, former financial advisor to the Egyptian government and first governor of the National Bank of Egypt (established in 1898).[46] Located south of Cairo, Ma'adi was prized for its proximity to Hilwan, which was known for its therapeutic sulphur baths. On the other hand, Heliopolis was developed by a rival business group led by Belgian transportation magnate Baron Empain, along with Boghos Nubar, chairman of the Cairo Water Company and son of former Prime Minister Nubar Pasha.[47] Marketed as a model, healthy suburb, Heliopolis sprawled across a vast desert area of 'Abbasiyya.

'Abbasiyya and Shubra attract a large number of residents during the boom decade. First developed in the mid-nineteenth century by Khedive 'Abbas I, 'Abbasiyya was largely an empty desert except for a few royal palaces, army barracks, and small pockets of poor housing where Nubian and Sudanese migrants had settled by the 1880s.[48] Similarly, Shubra was once an idyllic area sparsely populated by palaces and mansions. Unlike Ma'adi and Heliopolis, 'Abbasiyya and Shubra were developed incrementally by individual proprietors and small contractors. The two neighborhoods offered decent housing conditions and more affordable rents, attracting a large segment of the Egyptian middle class as well as European migrants from Greece, Italy, and Malta. This made them some of the city's most ethnically and religiously diverse quarters. While traditional neighborhoods were somewhat segregated along religious lines, 'Abbasiyya and Shubra drew a mix of Muslims, Christians, and Jews.

The construction boom ushered in a new generation of buildings that differed markedly from the old in terms of aesthetics, height, density, materials, and hygienic standards. The new buildings were designed with enhanced indoor plumbing and built-in sanitary amenities. Ma'adi and Heliopolis stand as particularly noteworthy examples of this change. For instance, Ma'adi's layout was deliberately optimized for the installation of water mains and electric cables. The suburb's managers issued detailed regulations specifying the building-to-plot ratio, height, and even the colors of walls and shutters, alongside requirements for indoor plumbing and cesspool construction.[49] Meanwhile, Heliopolis promised comprehensive municipal services. Its regulations included stringent requirements for ventilation, private bathrooms, and water and sewage tanks. The development company also conducted rigorous inspections of sanitary conditions, requiring owners to promptly address leaks and construction defects.[50]

The first showrooms for water and sanitary appliances were opened during the boom. In 1903, Max Steinauer, a Polish sanitary engineer and entrepreneur, established a shop in Azbakiyya where he showcased the latest products from English sanitary manufacturing. In the following years, his firm opened a large workshop in Bulaq and installed sanitary appliances in private homes, royal palaces, schools, hospitals, and prisons (Figure 20).[51] Another store, opened by the English firm Ratcliffe, specialized in hardware and sanitary appliances and was located near Cairo's main railway station.[52]

The Cairo Sewage Transport Company also prospered during the boom, particularly with increased reliance on waterproof cesspools. Since 1895, the company had focused on producing high-quality fertilizers. In 1899, it took over the less successful Société Générale Égyptienne des Engrais (General Egyptian Fertilizers Company), acquired additional land in Old Cairo and 'Abbasiyya, and broadened its product range by using animal carcasses from nearby slaughterhouses and quarantine stations. The company produced fertilizers such as "bone powder" and "pulverized blood," boasting its scientific approach to waste treatment and its ability to create products rich in nitrogen, phosphoric acid, and potash. The company's chairman, Henry Meyer, was a Swiss businessman and one of the founders of the Artesian Boring and Prospecting Company.[53]

MAX STEINAUER & CO.

1, 3, and 4. The Showrooms. 2. The Stores.

FIGURE 20. Max Steinauer's showroom. Wright and Cartwright, eds., *Twentieth Century Impressions of Egypt*, 361.

The building boom produced stunning effects not only on the physical landscape but also in the real estate market. Pierre Arminjon, professor at the Khedival School of Law and co-founder of the Society for Political Economy, Statistics and Legislation, sought to quantify the upward spiral of land values. He estimated that land prices in Cairo and Alexandria at least doubled between 1904 and 1907.[54] At an extreme end, the riverside area of Qasr al-'Aali Palace valued at £80,000 in 1902, sold for £700,000 in 1906. A small property initially worth £2,000 changed hands four times within a few years, ultimately selling for £32,000 in 1906.[55] It was not uncommon for urban properties to be bought and sold several times within a single year, leading to a significant increase in land values and living expenses, particularly rents.

The most frenzied phase of the bubble was characterized by rampant speculation in land, securities, and cotton futures—contracts to buy or sell a future crop at a predetermined price. In Cairo, some traded the securities of real estate companies on the stock exchange, while others took out bank loans to purchase empty plots in the suburbs, selling the properties at a profit before their first mortgage payment was even due. Land and stocks were passed from one hand to another like currency, each transaction inflating their perceived worth.[56] Some speculators genuinely believed that land and stock prices would rise indefinitely. Economic historian John Kenneth Galbraith analyzed this mindset as one of two common types in economic bubbles. The second type comprises those who believe that their unique insight allows them to "ride the bubble" until the precise moment to sell assets for maximum gain. The latter type is more realistic only in that they acknowledge the inevitability of a market crash.[57]

The Rawd al-Faraj Well Water Scheme

The building boom reshaped Cairo's built environment in a way that made it more amenable to private water consumption. As the holder of the monopoly on water supply, the Cairo Water Company was deeply invested in this urban transformation. The company also drew inspiration by the Tanta Water Company's success in extracting "artesian" water. In 1901, directors of the Cairo company began to consider drawing the city's supply from deep wells instead of the Nile. Over the following years, the company drilled several test wells in Rawd al-Faraj, a riverside neighborhood north of Cairo, to vet this approach. With full government backing, it was supplying exclusively well water in 1905.

While the project was primarily promoted on public health grounds, it was also praised for addressing other persistent issues with the water supply. One of the foremost proponents of well water was Paul Kauffmann, the bacteriological surveyor of the Cairo Water Company since 1894, and former head of the government's Bacteriological Laboratory. Kauffmann viewed deep wells as an opportunity to overhaul the water supply system and solve the stubborn problem of "green water."[58] The company had long struggled with filtration issues, particularly around mid-June when the Nile turned green from Aswan to Cairo and the water acquired an unpleasant taste. Kauffmann was among the first to demonstrate that the green coloration was caused by vast quantities of microscopic algae, likely originating upstream near Wadi Halfa in Sudan. Under his guidance, the company tested various methods to filter green water, including more frequent washing of the sand filter beds. However, these efforts failed to eliminate the disagreeable taste. Therefore, Kauffmann viewed well water as beneficial from both scientific and business perspectives. It was a naturally pure source with a consistent chemical and bacteriological composition, unlike the Nile, whose quality varied seasonally. From a business standpoint, it could potentially eliminate the need for water filtration and thus significantly reduce costs. Accordingly, Kauffmann advised the Cairo Water Company to switch to well water. Following his recommendation, the company built its first experimental well in 1901.

Two additional factors accelerated the adoption of an underground water supply in Cairo. First was the reappearance of cholera in 1902–03, which galvanized the government and the water company to secure a pure water supply. Second, there was widespread recognition that the existing system was unable to cope with Cairo's growing demands. With the proliferation of taller buildings, residents on upper floors began to complain of weak water pressure. The company's infrastructure had been originally designed for a cityscape dominated by two-story buildings, but the height of new constructions imposed fresh demands on the distribution system.

In 1903, the government and the Cairo Water Company finalized an agreement to "completely abolish the supply of Nile water in the city of Cairo and its suburbs, and to replace it with water from artesian wells."[59] Representing the company was Charles Beyerlé, vice chairman of the Board of Directors. Notably, the government retained the right to declare the new supply unsatisfactory if the water pressure was not sufficient to reach the upper floors, with the terrace of the Savoy Hotel—roughly equivalent to the sixth floor—serving as the benchmark.[60] Built in 1903 in the upscale Isma'iliyya neighborhood by hospitality magnate George Nungovich, the Savoy Hotel featured modern luxuries like an electric lift and private bathrooms with tubs, bidets, and ample running water.

On April 25, 1903, the company's shareholders gathered to ratify the new agreement with the government. During the meeting, Beyerlé spoke enthusiastically about the future, proclaiming that the company had uncovered vast quantities of high-quality water beneath Cairo. According to Beyerlé, the water was chemically pure and entirely free of germs. He also noted that the experimental well had yielded 5,000 cubic meters of water daily for nearly a year. On this basis, he claimed, underground water was "as constant as the movement of the Nile."[61] Following his presentation, the shareholders ratified the agreement, and the government signed it the same day.

Later in 1903, the company acquired land in Rawd al-Faraj to establish the new waterworks. Initially, two experimental wells were dug, each yielding an average of 4,000 cubic meters daily. Both the government and the company were confident that this underground supply was sufficient to meet the city's needs. As for quality, the water was found to be clear, odorless, and free of harmful bacteria. Based on these results, the company proceeded to sink twenty-two wells, each reaching depths of thirty to sixty meters. At the lower base of each well, a sieve made of three layers of wire gauze acted as a filter, straining the water from the surrounding soil.[62]

The water company expanded in lockstep with the building boom. In addition to securing the right to distribute water to several new suburbs, it repurposed older facilities to establish the Compagnie Frigorifique d'Égypte (Egyptian Refrigeration Company), specializing in ice manufacturing.[63] The company also resolved a longstanding dispute with the government regarding the unused gunpowder manufacturing site.[64] Starting in 1904, rubbish heaps were cleared from the site, and the land leveled, subdivided, and sold to contractors.[65] At the height of the boom in 1905, the company's shares—originally valued at 125 francs—reached a record 1,260 francs.[66] Capitalizing on this, the company quadrupled its shares in circulation from 68,000 to 272,000, greatly expanding its financial resources.[67] In accordance with the 1903 agreement, the company also constructed a large reservoir at the foot of al-Jabal al-Ahmar—part of the Muqattam plateau east of Cairo—leveraging the high altitude to increase water pressure in its network (cover image).

Analyzing Prosperity

Witnessing these dramatic changes in the urban landscape, the public was left in a state of shock—but they were far from silent. The spectacular rise in land values, rents, and living expenses sparked fierce debate in the press and scholarly circles. How should this economic phenomenon be understood? Was it real "prosperity," as the British regime claimed, or merely a "mushroom growth" destined to collapse? Was the building boom driven by raw speculation, or was it the natural result of population growth and demand for better housing? The answers varied greatly depending on one's perspective. In Lord Cromer's assessment, speculation was a problem, but it did not seriously threaten Egypt's economic development, which was grounded in his administration's sound financial and agricultural policies. Many aligned with this view, seeing Cairo's real estate market as a legitimate reflection of the demand for better housing in a city with abundant land, rising wealth, a growing population, and a stubbornly high mortality rate. Some boosters even claimed that Cairo and Alexandria were experiencing an influx of affluent European migrants at rates comparable to New York—another pointed example of how the speculative mania led many astray. Census officials ultimately rebuffed this claim as far-fetched. In contrast, anticolonial critics emphasized the profit-driven nature of the housing industry and its negative consequences for urban society.

Lebanese writer Jurji Zaydan analyzed the boom in his celebrated monthly journal *al-Hilal*.[68] In 1906, he published a two-part article titled "Egypt's Financial Boom: Illusory or Real?" where he argued that the current prosperity was built on a solid foundation. He attributed the rise in land values in Cairo and Alexandria to both increasing wealth and population, as well as the evolving needs of urban dwellers. Zaydan noted that comfortable Egyptians were spending substantial sums in search of better housing, driven by "the spread of refinement, the progress of taste, and the concern for health."[69] Despite widespread concerns that speculation had inflated prices excessively, he predicted that land values would continue to rise in the future.

However, Zaydan's confidence was partly based on a questionable analogy between Cairo and New York. He believed that Cairo had experienced a significant influx of rich migrants from both Europe and the Egyptian countryside. In his view, the demographic trends in Cairo resembled those of New York between 1860 and 1890, just before a massive population boom.[70] Zaydan reasoned that if current conditions persisted, Cairo would see astounding rates of demographic growth in the coming decades. Writing in 1906, a year before the government was set to announce the result of the new census, Zaydan estimated Cairo's population at 900,000, a sharp increase from the 570,000 recorded in the 1897 census. He described this as urbanization "at the speed of lightening," noting that it was a global phenomenon observable not only in New York but in cities more comparable to Cairo, like Johannesburg and Bombay.[71]

In a sequel article titled "Inflation, or the Rise of Prices," Zaydan offered a cautionary tale to those complaining about the rise in living expenses. The piece provided a brief history of famines in Cairo, drawing heavily from the chilling account of Abbasid historian ʻAbd al-Latif al-Baghdadi (1162–1231), who had witnessed a famine during one of his visits. Al-Baghdadi recounted that food shortages became so dire and prolonged that people resorted to eating cats and dogs before turning to cannibalism.[72] Zaydan's message was clear: while current inflation was troubling, it paled in comparison to the far graver hardships the city had endured in the past.

By 1907, the Cairo-New York comparison had gained traction among economic boosters, even though there were no figures to support it. In "Statistics and the Financial Boom," Zaydan doubled down on the comparison, despite the release of the preliminary results of the 1907 census, which estimated Cairo's population at 650,000—far below the 900,000 he had anticipated. In the article, Zaydan threw everything he could at the 1907 census, arguing that the results "went contrary to what everybody expected."[73] When the final results of the census were published, it became necessary to correct the imagined demography of Zaydan and others. C. C. Lewis, director of the Census Bureau, noted that while Cairo had experienced robust growth, especially for a largely non-industrial city, it was still incomparable to New York. He pointed out that the total number of migrants to Egypt over several decades "approximates the number of migrants to the city of New York alone during a single year, estimated at 200,000."[74]

Yacoub Artin reached similar conclusions to Zaydan, though without the exaggerated demographic figures. In *Essai sur les causes du renchérissement de la vie matérielle au Caire* (*Essay on the Causes of Increased Living Expenses in Cairo*), Artin argued that the rise in prices was a natural outcome of economic progress. He acknowledged the significant increase in land and rental values, as well as the varying rise in living costs. To him, it was understandable that the local press was calling for regulating the price of essential foodstuffs.[75] He reminded the readers that the Egyptian government had set prices until the mid-nineteenth century, a task traditionally overseen by the market inspector, who also monitored weights, measures, and food merchants.[76] However, Artin was convinced that "the fluctuations of supply and demand" were the true drivers of the current price increases.[77] The government could mitigate these fluctuations but could never fully control them.

According to Artin, land and rental values began to rise in the 1860s, with the expansion of the Egyptian capital under Khedive Isma'il. The commercial development of Azbakiyya drove up rental values, making them comparable to those in major European cities. This trend took an even sharper turn after 1897, fueled by the growing demand for better housing. Forty or fifty years ago, Artin recalled, it was common to lease a house without proper water amenities: "A stone pipe, or even one made of wood and sometimes brick, led household water into a bottomless cesspit, and that was all."[78] But today, he noted, "property owners are required to have finished houses, painted and wallpapered; well-maintained water pipes and sanitary sewers are also required."[79] Furthermore, the desire for healthier living conditions has driven the upper and middle classes to migrate from the city's old quarters. In Artin's final analysis, although the rise in land and rental values might have been somewhat excessive, he was confident that Egypt was on a path to material well-being.

The bust was already looming on the horizon when Alfred Eid published *La Fortune Immobilière de l'Égypte et sa Dette Hypothécaire: Propriété Rurale, Propriété Urbaine* (*Egypt's Landed Wealth and its Mortgage Debt: Rural Property, Urban Property*). Eid, the son of Georges Eid, the Syrian-born Belgian consul of Cairo, and an investor in several real estate companies, had studied medicine in France before returning to Egypt, where he founded two Arabic medical journals, *Tabib al-'A'ila* (*The Doctor at Home*) and *al-Tibb al-Hadith* (*Modern Medicine*).[80] In his book, he analyzed what he saw as an internal "exodus" of affluent residents from old neighborhoods to new areas offering better health, comfort, and privacy. He argued that the old quarters had become irreparable due to the deteriorating conditions of essential services, roads, and housing. With the introduction of tramways, residents left these areas en masse, moving to new districts that offered improved living conditions.[81] Therefore, Eid predicted that urban real estate would remain one of the most secure investments in the country. Revealing his preferred investment strategy, he said that land closer to the city center was not only abundant but also safer in the long term than properties in distant suburbs.

Writing after the crash, Pierre Arminjon, a law professor and judge at the Mixed Court, argued that distinguishing between urban and suburban lands was crucial for understanding the housing market. He noted that suburban land prices had been severely inflated by speculation and were now falling significantly. However, the situation was different for land within the city proper. In fact, Arminjon predicted that urban land values would either maintain their current levels or even rise slightly in the near future. He believed that land values followed the movement of rents, which continued to increase due to "the scarcity of hygienic, well-designed, and comfortable housing."[82] According to Arminjon, the search for better housing conditions is

> increasingly indispensable for the European or Europeanized population, whose numbers are growing so rapidly in Cairo and Alexandria. In all countries, this need is becoming more pressing, and families, especially in the middle class, are dedicating an increasingly larger proportion of their income to its satisfaction. In Cairo, only fifteen or twenty years ago, there were no other dwellings than the dark and damp hovels of the indigenous quarters.[83]

At the end, Arminjon believed that the rise and fall in the value of landed property was not disastrous. He was confident that large construction companies were backed by investors wealthy enough to weather the crisis.[84]

Ahmad Hilmi, editor and contributor of the daily newspaper *al-Liwa'*—a leading voice in the emerging nationalist movement—offered a pointed critique of the building industry.[85] As historian Aaron Jakes has shown, Hilmi excelled at deflating the British regime's official discourse, which he mockingly labeled "gilded speech" for its extensive use of figures and statistics that purportedly showcased Egypt's economic progress under colonial rule.[86] In his 1903 article "Urban Development of the Capital," Hilmi contended that soaring land values in Cairo could not be explained solely by increased wealth, population growth, or the demand for better housing.[87] Scrutinizing recent development trends, he argued that the chaotic nature of urban expansion was not a reflection of true needs but rather the interests of a "particular class" of businessmen investing where the highest profits could be found. With his characteristic eloquence, Hilmi described Cairo's growth under these conditions as "mobile urbanization" (*al-'umran al-mutanaqqil*). In the same article, Hilmi urged the government to address the needs of the neglected and overcrowded old quarters. However, some of his proposed solutions reflected a paternalistic attitude characteristic of the nationalist movement. For instance, the suggestion to "straighten" the winding streets of old neighborhoods revealed that Hilmi, like much of the Egyptian middle class, was already detached from traditional ways of living. Still somewhat influenced by the miasmatic theory of disease, his proposal lacked any input from the residents themselves.

Back to the Nile

While debates about housing, land values, and living costs raged on, Cairo's new water supply was also becoming mired in controversy. With the support of the government, the Cairo Water Company had switched the city's supply from the Nile to underground aquifers. After drilling several deep wells in Rawd al-Faraj, the company inaugurated the new supply in 1905. Despite the initial enthusiasm, complaints about water quality began to emerge. Moreover, the yield from the wells diminished, leading to the stark realization that there was no "hidden Nile" to be exploited. These issues sparked extensive debate in the press and among scientists and officials. After five tumultuous years, the company abandoned the supply and returned to the Nile.

Initially, government officials hailed the well water supply as a triumph in public health. In 1905, Horace Pinching—director of the Public Health Department from 1899 to 1907—lauded ongoing efforts to provide pure water to Egyptian towns as "the most important step taken up to the present by the government in the direction of improving the general hygienic conditions of the country."[88] The *Lancet* journal echoed this sentiment, asserting that "Cairo is now to be congratulated upon having for the first time a pure supply of drinking water."[89] Public health experts, who had long advocated for such improvements, shared in the enthusiasm. The government also planned to install several standpipes to provide well water freely to the poor, aiming to eliminate the reliance on remaining water carriers who distributed unfiltered Nile water.

However, objections to the new supply emerged as soon as it began reaching the city's taps. Residents complained about the water's color and odor, its suitability for cooking and washing clothes, and its adverse effects on hair.[90] Confronted with these issues, the government's initial response was to attribute the discontent to public resistance, claiming it was an irrational reaction stemming from a deep-seated preference for Nile water. In 1907, Lord Cromer reported to the Foreign Office that the controversy surrounding the well water was largely fueled by "imaginary properties the native population believe it to possess—one idea being that it produces sterility among their women."[91] The *Lancet* reported that "among the native population there was a superstition against drinking what they denominate "dead water." They will often prefer to drink extremely foul surface water to a pure water coming from underground where they say it is dead and buried."[92]

As complaints circulated widely among all segments of society, the government adjusted its response with a twofold approach. First, the government insisted that well water was healthy. In 1906, the Public Health Department published a report assuring the public that the new supply was germ-free and therefore safer than the Nile. Second, the same report stated that there were "no very pronounced differences" between

Nile water and well water.[93] According to the report, the water company had served well water to Bulaq's residents for nearly two years (1903–05) without complaints. "Only after it became public knowledge that the Water Company supplied well water instead of Nile water did different objections arise against the new supply," the report continued.[94] In other words, the government argued that the negative reception of well water was a matter of collective hysteria, triggered solely by its substitution for Nile water.

However, it turned out that there was an important difference between the two supplies. The same report admitted that well water was "hard," containing large amounts of dissolved minerals, whereas Nile water was "soft." Despite this, the report emphasized that the mineral content of well water was harmless and that it was still softer than that of major European cities. According to the report, the difference between the two supplies was detectable only by chemists in laboratories, not by ordinary consumers. The government had been aware of this issue since at least 1897, when Dr. Bitter noted that "the taste of the Egyptian people is somehow spoiled by the continued use of water from the Nile, which is extremely soft."[95] As the controversy raged on, the *British Medical Journal* suggested that the problem lay with "the delicate stomachs of some water-drinkers," rather than the chemical composition of the new supply.[96] Nevertheless, the public continued to avoid well water, leading one British official to quip, "you may take an Egyptian to the well water, but you cannot make him drink."[97]

In 1907, Lord Cromer left Egypt, and Eldon Gorst succeeded him as British consul-general. The transition occurred against the backdrop of the "Dinshaway incident," in which a pigeon hunting accident in Dinshaway, a village in the Minufiyya governorate, escalated into deadly clashes between British army officers and villagers. After summary trials, several villagers were publicly flogged, imprisoned with hard labor, or sentenced to death.[98] These events sparked a nationalist uproar and likely influenced the government's shifting stance on Cairo's water supply. In 1908, Gorst ordered an investigation to "determine what practical and reasonably economical methods can be adopted in order to convert Rod-el-Farag water into a high-class potable water, approximating in character to that of the Nile."[99] Ironically, despite previously attributing the issue to Egyptians' irrational preferences and their "spoiled" taste, the colonial administration now sought to modify well water to make it more palatable to the public.[100] Ultimately, however, the attempt to convert well water into something akin to Nile water failed.

As the controversy persisted, the government began to acknowledge the issue of water discoloration. Yellow and brown water led some to suspect sewage contamination, a claim categorically denied by the government. Instead, the Public Health Department argued that the discoloration might be linked to the water mains: as the Cairo Water Company increased water pressure, rust was being dislodged from old pipes. However, the government insisted, iron levels in the water were not hazardous.[101] By 1909, the Public Health Department discovered that elevated levels of iron and manganese in the water had led to the growth of a type of bacteria that produced black mold and exacerbated the discoloration. But still, the government maintained that the issue posed no threat to public health.[102]

At the same time, scientific research cast doubt on the very concept of "artesian" water. Austrian chemist Alfred Lucas criticized the assumption that there was no communication between surface water and groundwater below the clay stratum. Lucas had begun his career in Egypt as a water analyst, later becoming a celebrity forensic expert and member of the Tutankhamun excavation team.[103] In his 1908 book *Chemistry of the River Nile,* he argued that water drawn from deep wells was essentially Nile water that had infiltrated into lower geological strata, with the "hardness" resulting from minerals in the surrounding subsoil. While Lucas acknowledged that compact clay layers could be found in various locations, he found no evidence to suggest that there was one continuous layer running beneath the entire Nile Valley.[104] Instead, there may be several discontinuous or overlapping layers, with some locations, including Rawd al-Faraj, having no clay layer at all.[105] In other words, the notion of a single clay layer effectively separating polluted surface water from clean groundwater was not a scientific reality, but rather the invention of Egypt's economic boosters.

In 1909, the *Lancet* concluded that Cairo's well water was "no longer merely a sanitary and scientific problem," but one with "a political and sentimental aspect with which it is now necessary to reckon."[106] Alluding to the Dinshaway incident, the article suggested that the new supply should be discontinued: "Considering that the position of the British advisers to the Egyptian Government is fraught with many difficulties, and that occasions of acquiring popularity do not occur every day, it would be a pity to miss the present excellent opportunity of supporting a measure that would be so widely applauded."[107]

At the same time, the Cairo Water Company noticed a new, and more pressing, problem: the yield from the Rawd al-Faraj wells had diminished significantly, sometimes producing less than half of their original output. Investigations by government and company officials revealed that this problem was caused by buildup of siliceous material on the gauze of the intake pipe that had become cement-like and very difficult to remove.[108] Frustrated by this result, the company discontinued the new supply and went back to the Nile in 1910. Thus, the well water scheme ended in a debacle that would continue to haunt the British administration for years.

As the company reverted to the Nile, it implemented several measures to avoid the problems of the past. For instance, the intake site was relocated further upstream to allow for better supervision. Additionally, the Rawd al-Faraj facility was reconfigured to accommodate new water filters. The decision on which filtration system to use led to some more discussions, ending with the adoption of the "Jewell system" for rapid filtration.[109] Developed by the American Jewell Export Filter Company, this system was extensively marketed abroad, including in England, Germany, Austria, Holland, Russia, and Egypt.

The Boom's Aftermath

Typical of economic crises, the causes of the 1907 economic collapse were difficult to isolate and analyze. That year, markets tumbled across various parts of the world, including the United States, Chile, Japan, and Germany. American financial columnist Alexander Noyes concluded that the world had become an interconnected chain of credit relations.[110] A collapse in one link could therefore trigger aftershocks far beyond its geographical boundaries. The crisis exposed the increasing vulnerability of the Egyptian economy to the fluctuations of global financial markets.[111] In Cairo, it drove down land values and forced several building companies into liquidation. For the first time in a decade, construction and demolition slowed down. Foucart cynically remarked that the unfinished buildings and concrete foundations visible from his hotel room were, in a way, worthy of New York.[112] Another observer summed up what took place in Isma'iliyya: "They demolished palaces and villas and devastated the gardens without building anything. Instead of the projected New York, we only see ruins and rubble."[113] Another foreign observer noted how Shurba's main road had turned into "a hideous spectacle of grubbed-up trees and [building] foundations."[114]

The boom left a lasting imprint on the urban fabric. At the hands of real estate developers large and small, Cairo had grown rapidly without much coordination, leading to many irregularities and development at cross-purposes. Profit-driven development not only ignored the needs of the majority, as Hilmi had argued, but was also devoid of any larger vision for the city's future. The boom incentivized jerry-building—construction using cheap materials intended for quick profits—leading to the proliferation of indifferently arranged houses, blocks, and streets. In the following years, government officials began to grasp the extent of the damage, especially caused by small-scale developers. Whatever corrective measures the government decided upon, they would entail the additional costs of demolition and the modification of infrastructure.

Moreover, the boom strained the urban environment and generated unhealthy effects. Several observers noted that the rebuilt quarter of Isma'iliyya—the supposed showcase of modern urbanism—was not so healthy as claimed. Foucart disapproved of the pilling up of massive buildings, criticizing them from both aesthetic and public health perspectives. In his opinion, replacing palaces and gardens with densely packed six-story buildings has rendered the quarter "so obviously far removed from our modern concepts of air and light."[115] In 1902, Muhammad Wasfi, a teacher at the Khedival Arts and Crafts School, criticized deficient ventilation in "foreign-type" buildings, comparing it negatively to the superior aeration system of Cairo's traditional houses.[116] In 1909, the prominent liberal intellectual Ahmad Lutfi al-Sayyid estimated that only 5 percent of existing buildings met health standards. According to him, even the "modern homes" of Isma'iliyya were defective because they were built in the interest of profits more than hygiene.[117] However, he still believed they were better than houses in old quarters, which had become, "excuse me—microbial hotbeds."[118]

The logic of intensive land and water use backfired. In Isma'iliyya, increased levels of private water consumption led waterproof cesspools to fill up more quickly. The high cost of clearing these cesspools dissuaded many landlords from hiring the Cairo Sewage Transport Company, allowing "serious flooding" to occur.[119] Some landlords found themselves dedicating almost the entirety of their rental income to emptying cesspools, while others opted to simply cut off the water supply.[120] The *Lancet*'s commissioner pointed out that infiltration water contributed to this problem. In the spring of 1910, he visited Isma'iliyya to inspect the sanitary conditions of buildings. This is what he observed:

> In a narrow yard between two big blocks of buildings I was shown the flagstones in which were the openings of vast cesspools into which both these blocks drained. At times the cesspools overflowed into the yard and the stains on the walls showed how high this concentrated sewage might reach. The inhabitants had the cesspools emptied, but this was of no use. In a few days they were full again, not with soil from the adjoining blocks but with infiltrations from the surrounding land, and when the Nile rises the cesspools overflow into the yard.[121]

Therefore, "the atmosphere within these hot and lofty dwellings reeks of the effluvia from the soil below."[122] If these were the conditions in the heart of modern Cairo, then the building industry's promise of healthy housing could only be taken with a grain of salt.

Conclusion

Cairo's historians have shown that the building boom of 1897–1907 left an enduring mark on the city's fabric. This chapter has argued that the boom's effects went beyond spatial dimensions, extending to the fundamental act of using water, and in doing so, shaped what it meant to be a modern subject. The boom transformed the material conditions of everyday life by introducing residential architecture specifically designed for private water consumption. As Wilson Jacob, Nancy Reynolds, and Lucie

Ryzova have shown, this period saw the rise of the Egyptian middle class—the so-called *effendis*.[123] The *effendis* were marked not only by their education, profession, residence, clothes, shopping habits, and bodily comportment, but also by how much water they used and where. In other words, faucets, baths, flushing toilets, kitchen sinks, and waterproof cesspools contributed to the construction of modern subjectivity in Egypt.

As argued in this book's introduction, building water supply networks was not sufficient to initiate the global revolution in water consumption. Two crucial conditions remained: 1) older buildings needed to be replaced by others designed for private water usage, and 2) mechanized sewerage was required to drain the water consumed. Putting all these elements in place typically took several decades, as it involved considerable political, social, and economic challenges. In Cairo, the boom fulfilled the first of these conditions; the second, however, remained unresolved. As the next chapter will show, the Egyptian government reopened the Cairo drainage project in 1907, just as the bubble was beginning to burst. By undertaking this project, the British administration sought not only to address enduring criticism of its handling of public health, but also to remedy the problems resulting from the construction boom. In parallel with laying the sewer network, the government studied ways to mitigate the effects of Cairo's chaotic, profit-driven "mobile urbanization." As an outgrowth of these efforts, the British administration finalized an urban planning scheme that sought to regulate Cairo's growth for the next half-century.

FIVE

Sewers and Roads

Hydraulic Engineering and Urban Planning at the End of Empire

Between 1911 and 1920, the British colonial administration in Egypt developed an ambitious plan to address Cairo's problems. The plan aimed to remedy congestion, the lack of zoning regulations, and the shortage of affordable housing, while controlling the direction of urban growth over the next fifty years. By then, the government had acknowledged that the building boom of 1897–1907 had created more problems than previously anticipated. Rather than addressing the need for health and privacy, the unregulated real estate market had produced a housing crisis and a fragmented urban landscape lacking a coherent vision. However, as colonial officials were finalizing the details, mass demonstrations erupted in the streets, raising serious doubts about the immediate future of British rule in Egypt, let alone the next half century. Ultimately, the plan was swept aside by the political aftershocks of the 1919 revolution, which led to Egypt's conditional independence in 1922 and the establishment of an independent monarchy ruled by the descendants of Mehmed 'Ali Pasha.

The plan aligned with the global rise of the urban planning profession, whose appeal was based on the promise of not only improving cities but also alleviating social and political tensions, even quelling unrest in the colonies. For this reason, urban planning "was rapidly elevated to a high social purpose and its practitioners sought after in many parts of the [British] Empire," as historian Robert Home observed.[1] Between 1910 and 1920, British urban planners worked in India, Palestine, Egypt, Malaya, and South Africa—colonies that were simultaneously experiencing a wave of anticolonial mobilization and violence. In India, for instance, the 1912 assassination attempt on the British viceroy, Lord Hardinge, shook the colonial administration. In Egypt, political discontent intensified after the 1906 Dinshaway incident, peaked in 1910 with the assassination of Prime Minister Butrus Ghali, and culminated in the 1919 revolution.

Colonial urban planning in Egypt was explicitly framed in political terms. The Cairo plan was part of a broader town-planning program that included Alexandria, Tanta, and other provincial towns. One key figure in this program later admitted that town planning "was considered useful in creating public interest in the local politics of the 'village pump' and so to absorb activities which might otherwise be expanded in less desirable channels."[2] In other words, the colonial regime attempted to galvanize the public around issues of local, material well-being to distract from the clamor of national politics. While anticolonial critics increasingly called for constitutional governance and political representation, the British regime promoted a program of local "improvement" overseen by colonial officials and municipalities with limited authority.

This chapter explores the origins of the 1920 Cairo plan, arguing that it was an outgrowth of the sewage system project, which was finally implemented in 1915. After more than two decades of study and preparation, the government proceeded with the project, funding it directly from its coffers. At a cost of two million Egyptian pounds (later adjusted to 2.3 million), the project was the most expensive engineering venture ever undertaken in an Egyptian city. Notably, one-third of the total expenditure was earmarked for street repair, widening, and construction. Because the major streets were slated for excavation to install the sewer pipelines, the network's construction provided an opportunity to reconsider the city's street system and address related issues of traffic, circulation, and land use. In this sense, the project laid the institutional groundwork for analyzing urban space as a cohesive whole, introducing practices that eventually evolved into formal town planning.

As various studies have shown, the link between infrastructural engineering and urban planning was demonstrable in other cities as well.[3] Patrick Geddes, the Scottish sociologist and urban planner, was one of the experts advising on town planning in the colonies. In India, he criticized the way municipal engineers had led urban reconstruction efforts focused on extending water and sewer networks, building roads, and demolishing slums.[4] His intention was to show that town planning was an "art" that should be approached with a different mindset, more in tune with community needs and environmental considerations.[5] Nonetheless, his observations reveal that civil engineers were, in practice, the chief planners of British colonial cities.

The implementation of Cairo drainage led to several developments that are worth noting at the outset. First, it sparked significant discussions among experts and in the press. These experts debated its potential impact on water consumption patterns, with some cautioning that it should be designed to accommodate increased demand in the future. As seen in Chapter 2, sanitary engineers had been keenly aware of this issue for some time. Second, the government raised the house tax from 8.3 to 10 percent to fund the project—a measure that had been planned since 1892 (Chapter 3). Anticolonial voices campaigned against the tax hike, seeking to condition the increase on the creation of an elected municipality in Cairo. In doing so, they drew on the foundational political principle of no taxation without representation. Finally, because the Cairo drainage project was designed and managed by foreign engineers, it deepened the sense of marginalization experienced by Egyptian engineers since the onset of British rule, breeding resentment and political militancy.

Sewers at Last

In 1904, Britain and France signed the Entente Cordiale, an agreement that eased their imperial rivalry in the Middle East. Under this arrangement, Egypt was recognized as being within the British sphere of influence, while Morocco fell under French control. This agreement created favorable financial and political conditions for the Egyptian government to revisit the long-delayed issue of Cairo's drainage and to finally implement the project. First, it unlocked substantial money that had been accumulating in the Egyptian reserve fund, over which France had previously held veto power. With the agreement in place, the British colonial administration gained access to six million Egyptian pounds—nearly half of Egypt's annual revenue—without French interference. These funds were used to cover the entire initial capital expenditure of the sewage system project.[6] Second, the Entente Cordiale paved the way for increasing Cairo's house tax from 8.3 to 10 percent, a measure that could then pass without French opposition. The additional revenue, estimated at L.E. 40,000 to L.E. 50,000, would help finance the system's maintenance and operating costs.[7] Finally, the British administration was no longer compelled to accommodate French business interests, as had been the case in the past (Chapters 2 and 3). Freed from these constraints, the government could now appoint a British engineer, grant him full control over the project, and award the largest contracts to British firms. This is precisely what occurred.

In 1906, the Egyptian government approached British engineer Charles Carkeet James to design the Cairo drainage project.[8] James had begun his career as a trainee on various railway, water, and drainage projects in Cornwall, England. He later moved to India, where he worked for the Bombay Municipality and became involved in water and sanitation initiatives. In 1897, he oversaw Bombay's drainage scheme and served on the Bombay Plague Committee.[9] In 1902, he published *Oriental Drainage*, later expanded and retitled *Drainage Problems of the East*—a textbook outlining the principles of sanitary engineering in "eastern cities."[10] These experiences and publications made James particularly appealing to the Public Works Ministry, whose officials believed that Cairo required a different approach from European cities due to its unique climate and the prevalent use of collective water facilities such as mosques, baths, and fountains.

The government provided James with two decades' worth of reports and documents generated by previous studies, and he was given two years to review these materials, survey potential sites, and draft a preliminary plan. During this period, James collaborated closely with William Garstin, advisor to the Public Works Ministry, submitting several proposals before securing Garstin's approval.[11] The main question was where to place the pumping station and the sewage farm. Garstin mandated that they must be situated as far away as possible from inhabited or soon-to-be-developed areas. At the same time, if placed too far, the costs would increase on account of having to lay the sewers deeper. Ultimately, the two agreed to place the pumping station close to the village of Kafr al-Jamus (now 'Ayn Shams) and to build the sewage farm in Khanka, twenty-six kilometers northeast of Cairo.[12]

Early on, the Public Works Ministry decided to undertake extensive road work in parallel with the sewage system project. Because laying pipes required digging trenches and reconstructing pavement, the ministry recognized a significant opportunity to improve the city's thoroughfares. During the previous decade, the government had paved the streets of upscale neighborhoods with macadam and asphalt to accommodate increasing traffic.[13] The intention was to extend this effort to poorer quarters while laying the sewer pipelines. Responsibility for this side of the project fell to the Tanzim Department, which managed all road-related matters.[14] The Tanzim, a part of the Public Works Ministry, was one of three separate branches of the central government responsible for managing Cairo's affairs, alongside the Public Health Department of the Interior Ministry and the Cairo Governorate.

The design of the sewage system was based on three key decisions. First, James determined that the system could not rely solely on gravity.[15] Arriving in Cairo during the height of the real estate boom, he thought it was impossible to drain the sprawling city using a single mechanism. To address this, he divided Cairo into two main areas, each employing a different method. First was the "sectional" area, which was divided into small sections, each equipped with an ejector for pumping sewage toward the main collector. The sectional area encompassed the core urban districts, including Bulaq, Shubra, and Old Cairo (Figure 21). Second was the "gravitational" area, which relied primarily on gravity. James designed this feature to make use of the increased depth of the main collector as it approached the main pumping station. The gravitational

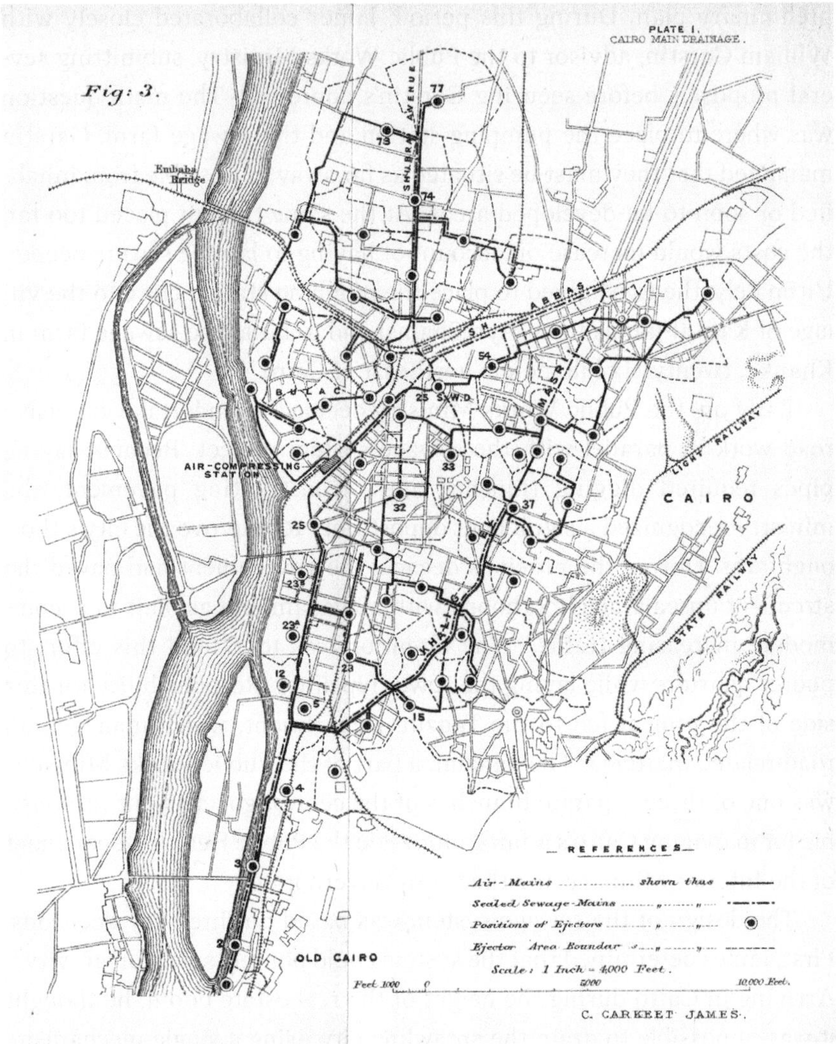

FIGURE 21. The sectional area of the sewage system scheme. James, "The Main Drainage of Cairo," Extract from Plate 1.

area included the city's northeastern suburbs—Qubba, Zaytun, Matariyya, and Heliopolis.

The second major decision was to drain surface water separately from domestic wastewater. As James confessed, this was an unusual decision. According to standard practice, the two were typically drained together in contexts where rainfall was not a year-round issue, such as Cairo.

However, two key factors swayed his opinion. First, since 1904, the government had installed stormwater drains in the city's wealthy districts.[16] James argued that demolishing these drains would be too expensive, therefore, it was less costly to expand on the existing infrastructure. Second, James argued that Cairo's occasional but typically disruptive rainfall presented an unusual challenge (Figure 22). He believed that calculating the system's total capacity based on rare instances of heavy rainfall was unwarranted, even "extravagant," leading to much higher costs.[17] Accordingly, James planned to expand the surface drainage network. Under normal conditions of little or no rainfall, surface water would be led into the sewers; during rain events, it would be redirected to the Nile and the Isma'iliyya Canal. In this way, James sought to separate dirty street washings from surface water during heavy rainfall when it was sufficiently diluted. As will be seen, this decision would later face scrutiny.

Finally, James turned to the thorny issue of estimating future water consumption, which had led to so much argumentation in the past (Chapter 2). James designed a system intended to serve 940,000 people, a figure

FIGURE 22. Marasina Street (in Sayyida Zaynab) after heavy rainfall, 1909. Ministry of Public Works, *Sewerage of Cairo, First Section, Report and Drawings*, unnumbered page.

projected for Cairo—then a city of 644,000—by 1932. Like other engineers who previously studied the issue, James had to take the uneven geography of water distribution into account. By then, about 15 percent of the city's buildings were connected to the water company, and they were mostly concentrated in the areas built since 1867. In the rest of the city, residents still relied heavily on mosques, baths, and fountains, while buying what they could from the company's standpipes.

Therefore, James divided the system's users into two classes, each with a distinct water allowance.[18] He provisioned what he called "upper class" users—100,000 Europeans and Egyptians living in new areas—with 150 liters per capita per day. As for "lower class" users—544,000 inhabitants of old areas—James allocated an allowance of forty-five liters per capita per day (ten liters at public toilets, twenty at mosques and baths, and fifteen at home). According to his plan, the second class of users could connect their houses to public sewers through low-capacity gullies. However, he cautioned that this process would be slow, as the financial responsibility for the sewer connection lay with the property owners. Ultimately, he believed he had made a "liberal allowance" for the second class of users, considering the expensive rate of the water company: "The majority of the population buy all the water they require for domestic purposes from the 'bornes fontaines' [standpipes] and doubtless buy as little as possible."[19] However, as will be seen, several contemporary experts found this figure insufficient to accommodate the expected rise in water consumption.

In Search of Municipality

As soon as the government accepted James's proposal and authorized its implementation in 1909, it initiated steps to raise the house tax from 8.3 to 10 percent to cover the costs of maintenance and operation. However, the tax hike quickly became a rallying point for the critics of British rule, who condemned the imposition of taxes in the absence of a municipality or other institutions of political representation. During the previous decades, the British regime had consistently opposed the creation of a Cairo municipality, preferring to govern the city through the central administration. Lord Cromer, in particular, feared that local self-governance in the Egyptian capital could become a "stepping stone" for full independence.[20] In this way, the house tax unwittingly became part of the struggle for constitutionalism and representative government in British colonial Egypt.

Throughout British rule, Cairo was an exception to other cities. The regime authorized the creation of municipal bodies in Alexandria, Tanta, Mansura, and other provincial towns. But not in the capital.[21] Established in the 1890s, the Alexandria Municipality represented the most robust form of local self-government in Egypt. It was partly elected, financially autonomous, and had legal standing. Its responsibilities included collecting taxes and administering funds for public services such as building maintenance, roads, water supply, lighting, waste management, and drainage. However, unlike its counterparts in Europe, the Alexandria Municipality had no jurisdiction over the administration of justice or policing. Additionally, any municipal taxes had to be approved by all foreign Powers with interests in Egypt. Meanwhile, the composition of the council ensured that it represented the upper class and foreign communities. Fourteen of the twenty-eight council members were government appointees, while the rest consisted of representatives of export and import merchants and large property owners. The conditions of candidacy and franchise restrictions guaranteed that members were drawn exclusively from a narrow class of male urban elites. Historian Michael Reimer estimates that, around 1895, only 1 percent of Alexandria's population had the right to vote.[22]

By 1907, Egyptian nationalism had grown into a mass movement, attracting journalists, lawyers, students, doctors, engineers, and workers to the cause of political independence and self-determination.[23] While the Dinshaway incident played a pivotal role in the rise of the anticolonial struggle, other regional and international events contributed significantly. A global wave of revolutions—in Iran and Russia in 1905, the Ottoman Empire in 1908, Mexico in 1910, and China in 1911—called for the establishment of constitutional regimes. Despite their differences, these uprisings heralded constitutionalism as a solution to various social and political problems.[24] Egyptian nationalists closely followed these developments, particularly in the Ottoman center. They began forming political parties, such as al-Hizb al-Watani and Hizb al-Umma. They also began to organize in professional associations, most importantly the lawyers' syndicate. The years 1907–1910 were marked by an uptick in labor strikes, student demonstrations, and political violence.[25]

The lack of a Cairo municipality mirrored the larger absence of political self-determination, even where municipal bodies already existed. The nationalist journalist Ahmad Hilmi was among those who called frequently for local self-governance in the capital, but he was also quick to reject the Alexandria Municipality as a viable model because it excluded the vast majority of Egyptians.[26] In Hilmi's view, there was no point in establishing a municipal body in Cairo unless it was popularly elected and composed of an indigenous majority. He argued that this was the only meaningful measure capable of addressing the city's needs.

The call for a Cairo municipality emerged from various quarters, including those not typically associated with political radicalism. Among these was the General Assembly, a consultative body established by the colonial regime in 1883 to offer non-binding advice on legislative matters. Composed of Egyptian notables and landowners, the Assembly had functioned largely as a rubber-stamp institution for much of its history. However, it began to adopt a more confrontational stance during this period. In 1904, one of the Assembly members, 'Abd al-Rahim al-Dimirdash—a wealthy leader of a Sufi order—proposed the creation of an independent Cairo municipality.[27] However, the colonial government swiftly rejected the proposal, citing logistical and legal difficulties. In defiance, the Assembly appointed an expert to study the matter, hiring Toussaint Caneri, a former employee of the Cairo Tanzim Department with twenty-two years of experience. In 1905, Caneri issued a report—*Cairo City: Its Present and Future—Essay on the Creation of a Municipality*—in which he argued that a municipality was not only absolutely necessary to ensure the health and beauty of Cairo, but also financially feasible, with an expected revenue nearly five times higher than what the Tanzim Department typically allocated to the city. Moreover, Caneri addressed the challenge of taxing foreigners, pointing out that Alexandria had already proven it was possible. He expected the European Powers to cooperate for the sake of improving conditions for their resident communities.[28]

After the Dinshaway incident, the General Assembly renewed its call to create a Cairo municipality. This time, however, it was one of several bold demands put forward by a group of Egyptian notables, led by al-Dimirdash and Hasan Madkur, a prominent Cairo merchant. Assembly members pressed the government on the high rates for water, gas, and electricity—utilities controlled by corporate monopolies with a large share of foreign capital.[29] Even more daringly, the Assembly called for the future establishment of an Egyptian parliament. In the meantime, they proposed that the Assembly's powers be expanded to cover "everything related to public wealth and matters of general administration."[30] The suggestion to take full control of Egypt's finances caused a stir in government and business circles.[31] Although Lord Cromer quickly rejected these ideas, it was evident that the political climate was shifting. Cromer left Egypt in 1907, as public anger over the Dinshaway incident continued to escalate.

Eldon Gorst, the new British agent and consul-general, initially sought to ease some of the pressure by revisiting the question of a Cairo municipality. In his 1908 annual report, Gorst noted that the existing system—where Cairo was managed by three separate branches of the central government—was "conducive neither to economy nor efficiency."[32] He proposed that the city be "managed by one central body on which the different Departments interested, and also the inhabitants, were represented."[33] However, the brief period during which the colonial regime appeared willing to offer small concessions passed swiftly. Shortly afterwards, the government suggested creating a municipal body consisting entirely of government-appointed or -nominated delegates.[34] Thus, while the regime recognized the need to centralize Cairo's administration, it sought to avoid risky experiments in local political representation at a time when the call for full political independence was rapidly gaining momentum.

In this fraught context, the government sought to raise the Cairo house tax without making the situation more volatile. For the first time, passing a measure through the foreign Powers seemed easy in comparison to passing it locally. To ease the measure, the British administration decided to involve the General Assembly. This was a largely symbolic gesture aimed at giving the impression of public discussion on the tax hike, and thereby garnering a modicum of political legitimacy. At the same time, this move was carefully calculated to neutralize the Assembly by putting forth a request that could not be reasonably denied given the expected health benefits of the sewage system project. As the Assembly looked into the matter, nationalist newspapers like *al-Mu'ayyad* urged it to delay approving the tax hike until a Cairo municipality was created. However, these calls faced an uphill battle.[35] Ultimately, in February 1909, the Assembly agreed to the tax increase, but on the "condition" that the government reconsider the creation of a municipality.[36] Of course, there was no way to enforce this condition, and the issue was indefinitely postponed.

At the same time, the General Assembly became entangled in another problem with greater implications for Egyptian national politics. In 1909, the Suez Canal Company sought to extend its concession by an additional forty years (from 1968 to 2008). In exchange, the company offered the government four million Egyptian pounds and a profit share. Seeking to replicate what happened with the house tax, the British administration sought to legitimize the deal by passing it through the General Assembly. However, whereas the need for a drainage system could be felt by everyone involved, extending the Suez Canal concession was mainly slated to benefit the company's shareholders. By then the company had gained a reputation among anticolonial critics as one of the purest expressions of the exploitation of public resources for private gain. Muhammad Farid, a prominent leader of the nationalist movement, condemned the agreement as extremely unfair, contrasting the privileges enjoyed by foreign monopolies in Egypt with trends in Europe, where governments were breaking up railway and public service monopolies.[37] Another vocal opponent of the proposal was the journalist Amin al-Rafi'i, who argued that the government would lose significant revenue by extending the concession.[38] As the founder of two leading nationalist newspapers—*al-Sha'b* and *al-'Alam*—Rafi'i was able to reach a broad audience and played a pivotal role in the General Assembly's decision to reject the deal.[39] Faced with public outcry, the British administration and the Suez Canal Company dropped the idea.

Egyptian Engineers in the National Movement

Egyptian engineers were part of the movement calling for Egypt's independence from Britain. The assassination of Prime Minister Butrus Ghali provides a revealing illustration of this connection. Ghali was shot on

February 21, 1910, in broad daylight as he was boarding his carriage outside the Ministry of Justice. The assassin was a young pharmacist by the name of Ibrahim al-Wardani, who cited several reasons for his act, including Ghali's involvement in the Dinshaway trials and the plan to extend the Suez Canal concession. Subsequent police reports revealed that Wardani had mingled with Russian revolutionaries and anarchists during his studies in Lausanne, Switzerland. Upon returning to Egypt, he became an active participant in Cairo's labor syndicates and the night schools organized by nationalists in working-class neighborhoods. In addition, police investigations revealed that Wardani was a member of an underground group known as the Society of Mutual Brotherhood.[40] As it turned out, several other members of this group were engineers.

The assassination plunged Egypt into political turmoil, prompting the government to crack down on nationalists and initiate a witch-hunt for clandestine political activities. Wardani was swiftly charged, tried, and sentenced to death that same year. A police inquiry revealed that the Society of Mutual Brotherhood had been founded in 1905 by students at *al-Muhandiskhana* (the School of Engineering), with Wardani joining in 1908. According to the police charges, Wardani was recommended for membership specifically for his radical views and skills as a pharmacist knowledgeable in preparing dynamite and poisons. As the government detained other members of the society, it emerged that three of them were current employees of the Public Works Ministry: 'Ali Murad, an irrigation engineer in Fayyum; Mahmud Anis, an irrigation engineer in Minya; and 'Abd al-'Aziz Riff'at, a municipal engineer at the Cairo Tanzim Department.[41] The three represented a cohort of Egyptian engineers hired since 1903 to fill junior posts created by ongoing irrigation projects and rapid urbanization.[42] As previously shown, the higher echelons of the public works administration were dominated by British engineers. Egyptians found themselves silenced, underpaid, crowded out, and relegated to less significant roles. Evidently, these conditions fostered a sense of militancy among Egyptian engineers.

Little is known of what became of the engineers arrested with Wardani after their initial detention. After spending a month in custody, they were released and fired from their government positions. The newspaper *al-Sha'b* protested their dismissal as punitive, asserting that they had been targeted for their political activism and for communicating with the press, without any proof that they were actually involved in the murder.[43] However, fifty years later, one of these engineers—'Abd al-'Aziz Riff'at—spoke out in an interview with the Egyptian weekly *Akhir Sa'a* (Figure 23).[44]

Riff'at revealed that the three engineers fled Egypt upon learning that the government planned to rearrest and try them. He moved to Istanbul, where he lived and worked until 1960. In the interview, Riff'at recounted the formation of their group at the School of Engineering and their conclusion that the Prime Minister was a "traitor," especially in light of the proposal to extend the Suez Canal concession. He vividly described his time in custody, his interactions with lawyers, and his dismissal from government by C. Haswell, the director-general of the Tanzim Department from 1909 to 1921. Titled "Why I Fled My Country!," the interview ended on a personal note. Asked if he had a wish at his advanced age, Riff'at revealed that he was ill and wished to die in Cairo. When he fled, he carried an Ottoman travel document, as Egypt was still nominally part of the Ottoman Empire. Later, he acquired Turkish nationality but never became Egyptian on paper. With his Turkish passport, he could only stay in Cairo for short visits and had to return to Istanbul to renew his visa.

Under Construction

Amidst this atmosphere of political turmoil, the government began to build the Cairo drainage network late in 1909. In line with British policy, the project had hitherto been considered a matter of government secrecy, with no information released to the press—a strategy aimed at minimizing criticism of the colonial administration. However, as the government began to dig the trenches, details of the project emerged and came under public scrutiny. An Egyptian whistleblower flagged several issues, including flawed design, lack of expert oversight, and favoritism toward British companies. Other observers criticized the project's underlying assumptions, arguing that James had not adequately accounted for increased water usage in the future. The very notion of treating "eastern cities" differently from their European counterparts was called into question.

لماذا هربت من بلدى !

■ متهم فى قضية بطرس غالى يذيع السر بعد ٥٠ سنة ..

تحقيق يكتبه : جميل عارف

تكلم هـــذا الرجل بعد ٥٠ ســـنة بالضبط . ان اسمه عبد العزيز رفعت، وكان واحدا من سبعة شـــبان قبض عليهم فى سنة ١٩١٠ بتهمة الاشتراك مع ابراهيم الوردانى فى حادث اغتيـــال بطرس غالى ((باشا)) رئيس مجلس النظار .. اى رئيس مجلس الوزراء ..

انها قصة تاريخية قديمة .. واثناء التحقيق فشلت جميع المحاولات لاقناع الشبان السبعة بالاعتراف ..

اما الوردانى فقد وقف امام المحكمة ليقول باعلى صوته : انا القاتل .. لقد قتلته وحدى لانه وافق على ان يمد امتياز شركة قناة السويس لمدة اربعين سنة !..

واعدم الوردانى شنقا .. اما الشبان السبعة فقد امر ((قاضى الاحالة)) .. وكان اسمه متولى غنيم بالافراج عنهم فورا لعدم ثبوت الادلة ضدهم .

واراد الانجليز ان يستانفوا الحكم .. ولكن فجاة اختفى الشبان الســـبعة .. لقد هربوا من ((مصر)) .. واحيل قاضى الاحالة على التقاعد .

وعاش عبد العزيز رفعت فى تركيـــا خمسين سنة كاملة ثم عاد الى القـــاهرة منذ عدة اسابيع ..

انه الوحيـــد ((الحى)) من الشبان السبعة . لقد ماتوا جميعا .. وفى هذا التحقيق السريع اخـــذ يروى ((لآخر ساعة)) حقائق مثيرة عن تفاصيل المؤامرة التى انكر اشتراكه فيها عندما قبض عليه الانجليز فى سنة ١٩١٠ .

عبد العزيز رفعت .. عمره الآن ٧٥ سنة وسالته عن القاهرة فى سنة ١٩١٠ فيقول لك انهم كانوا يركبون البغال والحمير .. وكان الترام موجودا وتذكرته بـ ٦ مليمات واول سيارة ظهرت فى سنة ١٩٠٥ وكانت لوالدى الخديوى عباس حلمى ! .

فى هذا البيت فى حارة الدمالشة وضعنا الخطة لاغتيال بطرس غالى !

FIGURE 23. Interview with ʻAbd al-ʻAziz Riffʻat. Jamil ʻArif, "Limadha Haribt min Baladi!" *Akhir Saʻa*, March 16, 1960.

In fact, municipal experts noted that water systems transformed cities in ways that transcended the simplistic distinction between East and West.

Building the sewage network was a complicated process that took six years (1909–15) and demanded close collaboration between the Tanzim Department, headed by Haswell, and the newly created Drainage Department, led by Charles Carkeet James. Work on the systems' main components—the main collector, the pumping station, and the sewage farm—posed no serious challenges since they were built in open spaces and on the city's outer fringes. In contrast, digging street trenches had to proceed gradually to avoid the unpleasant situation of having the entire city under excavation. Therefore, the main streets were sewered between 1910 and 1915, while the rest of the network was gradually built over the following three decades.

In 1909–10, James prepared a public tender file containing detailed information, plans, and timetables for each component of the system.[45] Firms wishing to apply for contracts were given three months to prepare their proposals. The first contract was for the Engineers' Quarters, to be built in Isma'iliyya.[46] Another contract covered the main collector, which was to be constructed using reinforced concrete. A separate contract for the rising main was put out for alternative tenders, allowing bidders to choose from three different construction materials: cast iron, steel, or reinforced concrete. Additional contracts were prepared for the surface water drainage scheme, pumping machines, and ejectors for the sectional areas. Within a few years, Hughes & Lancaster of London secured the largest contracts, with a total value of L.E. 787,300. Two other contractors trailed behind: the French Société des Grands Travaux de Marseille, with contracts worth L.E. 90,600, and the local Italian businessman Samuel Sornaga, who secured a L.E. 100,000 contract for brick supply.[47]

As construction proceeded, the Public Works Ministry documented the work in progress with a series of photographs (Figures 24–27). One photograph shows a group of workers building what appears to be the main collector, with William Garstin possibly posing in the background. The other three photographs were likely taken in al-Muski—a commercial center and a gathering point for Cairenes from all walks of life. This location can be inferred from the shop signs and the distinctive shape of the Mixed Court visible in the background of one photo. The images

FIGURE 24. The main collector at Ghamra under construction. Public Works Ministry, *Sewerage of Cairo, Photos of the Work in Progress, 1909–1910.*

capture engineers from the Tanzim and Drainage Departments supervising the work, alongside workers transporting soil, digging trenches, or taking a break.

In April and May 1910, a series of articles appeared in the pages of the nationalist newspapers *al-Sha'b* and *al-'Alam* providing a critical review of the project. Signed by an unnamed "Observer," the four articles argued that the project was flawed on both technical and hygienic grounds, and problematic from a public management perspective.[48] While the author's identity remains unknown, the articles demonstrate a command of technical knowledge, with multiple references to engineering authorities, manuals, and journals. Therefore, it is reasonable to infer that Observer was an engineer who sought to vent under the cover of anonymity. Given the platforms, it is likely that Observer was a nationalist or at least a sympathizer with the anticolonial movement. Meanwhile, the level of detail and information presented in the articles suggests that Observer was most likely an Egyptian whistleblower from government circles. Curiously, the timing of the articles coincided exactly with the dismissal of members of the Society of Mutual Brotherhood from the Public Works

FIGURE 25. Excavation in Muski. Public Works Ministry, *Sewerage of Cairo, Photos of the Work in Progress, 1909–1910.*

FIGURE 26. Excavation in Muski, with the Mixed Courts in the background. Public Works Ministry, *Sewerage of Cairo, Photos of the Work in Progress, 1909–1910.*

FIGURE 27. Excavation in Muski. Public Works Ministry, *Sewerage of Cairo, Photos of the Work in Progress, 1909–1910.*

Ministry: the three engineers were fired in mid-April, just as Observer's articles began to appear at the end of the month.[49] As an employee of the Tanzim Department, 'Abd al-'Aziz Riff'at was in fact very close to the sewage system project, with potential access to confidential government information.

Observer's first line of argument was that a separate system for draining surface water was unnecessary, resulting in wasteful expenses. He pointed out that Cairo's climate did not warrant the construction of a separate network dedicated solely to managing rainfall and street washings. Such a system was necessary only in cities where heavy rainfall was a constant problem. Instead, Observer maintained, Cairo would be best served by a combined system in which both surface water and household sewage were collected in the same pipes from the outset. Relatedly, Observer indirectly questioned the system's capacity to accommodate the city's future growth. If the system were sufficiently large, then it should be able to handle Cairo's occasional downpours. If it was not large enough, then it was a waste of public money, because the sewers "cannot be changed every year."[50] Moreover, Observer criticized the design from a public health perspective, warning that dumping surface water into the Nile posed significant risks for communities still dependent on the river for their water supply. Citing the British Royal Commission on Sewage Disposal, Observer indicated that surface water was always susceptible to bacteriological contamination, even when heavily diluted by rainwater.

Overall, Observer condemned the fact that the project went into execution without independent examination or public oversight. One result of this situation, he revealed, was that British companies were awarded the largest contracts. This point garnered the attention of the General Assembly, prompting its members to demand answers from the Minister of Public Works.[51] Finally, Observer suggested that James was "blindly imitating" the lessons he had learned in India.[52] Belittling James's self-styled expertise in "oriental drainage," the Egyptian engineer raised the possibility that "the East" was not one and the same.

The sewage system project came up at the Egyptian nationalist conference of 1910. Held in Brussels, the conference aimed to advocate for Egyptian independence to a European audience. Dr. Mahgub Thabit delivered a speech focusing on the status of public health in Egypt.[53] A French-educated physician, Thabit was an influential figure in Egyptian politics, particularly within the labor movement. In his address, Thabit presented health statistics, addressed the well water supply fiasco (Chapter 4), and critiqued Cairo's drainage scheme. He lamented the absence of a municipal body to oversee the project, echoing Observer's criticisms of the design. Furthermore, Thabit underscored that British contractors had secured the largest share of the project, subtly suggesting that continental European companies were being systematically excluded from the British-dominated Egyptian market.

The Cairo drainage project also received critical commentary in the *Lancet*. Writing for the distinguished medical journal was Adolphe Smith, a British socialist journalist and sanitary reformer.[54] While Smith was relieved that the system was finally being built, he criticized what he considered a significant flaw. Smith argued that the design was predicated on the assumption that disparities between the two classes of users would largely persist in the future. The most glaring manifestation of this was the idea that the second class of users—the majority of the city's residents—would continue to rely on mosque latrines and public baths, consuming forty-five liters per capita per day even after the system was installed. To Smith, this assumption conveyed a "pessimistic" and "hopeless" tone.[55] Conversely, Smith argued that the system's introduction would encourage the use of home toilets and incentivize domestic water consumption, gradually narrowing the disparity between the two classes. In other words, the second class of users would come to use a lot more water at home than James had anticipated. Relatedly, Smith challenged common platitudes about the Egyptians' readiness for the new technology, proclaiming that "the British people were just as ignorant as the Egyptians" regarding the use of faucets, drains, and flush toilets when these innovations were first introduced.[56]

One year into implementation, the project's design was significantly modified. In 1910, the Cairo Water Company reverted to the Nile as the source of water supply, ending five years of public controversy (Chapter 4). The decision forced the government to rethink the surface drainage scheme even after considerable work had been done. The idea of dumping surface water in the Nile, even after heavy rainfall, was no

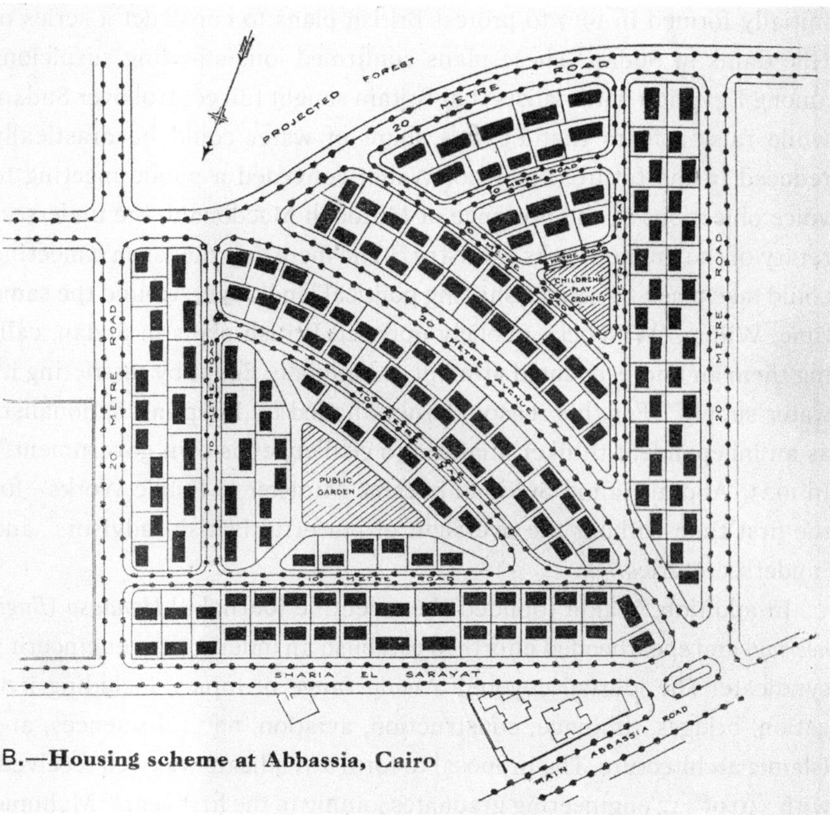

FIGURE 33. 'Abbasiyya housing scheme. Haswell, "Town Planning and Housing in Cairo," 258.

of the highest-ranking Egyptian engineers in government.[89] Sami's stated goal was to create an organization that could represent the country's professional engineers. In practice, however, the society primarily attracted elite members, with its total membership remaining under one hundred.[90] Notable figures included Ahmad 'Umar, the successor to Haswell as Tanzim Director-General, and Muhammad 'Irfan, who became head of the Main Drainage Department. The society gained official recognition from King Fu'ad (r. 1922–1936) in December 1922.

The second group, led by irrigation engineer 'Abdalla Wahbi, included engineer Ibrahim Zaki and the tireless doctor Mahgub Thabit. This group initially formed in 1919 to protest British plans to construct a series of Nile dams in Sudan. These plans confirmed long-standing suspicions among Egyptian nationalists that Britain sought full control over Sudan, while raising fears that Egypt's share of water could be drastically reduced in the future.[91] In 1920, Wahbi convened a public meeting to voice objections in the presence of Murdoch Macdonald, the undersecretary of the Public Works Ministry.[92] The mere fact that such a meeting could take place signaled a shifting political landscape. Around the same time, William Willcocks publicly opposed British plans in Sudan, calling them an unprecedented attempt to humiliate Egypt by restricting its water supply.[93] For this reason, he was hailed by Egyptian nationalists as an independent thinker unafraid to challenge his own government.[94] In 1923, 'Abd al-Hamid Suliman became Minister of Public Works—for the first time without the oversight of powerful British "advisors" and "undersecretaries."

In addition, Wahbi founded the scientific journal *al-Handasa* (*Engineering*) and spearheaded efforts to establish an independent engineering syndicate. The journal covered a wide range of topics, including irrigation, bridges, drainage, construction, aviation, natural sciences, and Islamic architecture. The proposal to form a syndicate was well received, with 340 of 442 engineering graduates joining in the first year.[95] Mahmud Fahmi, one of Egypt's most senior engineers, was chosen as the provisional head. Fahmi had enjoyed a distinguished career, having served as chief engineer of the Charitable Endowments Ministry and contributed to the Cairo drainage scheme of 1892 (Chapter 3). However, King Fu'ad refused to recognize the syndicate. In fact, official recognition would not be achieved until 1946.[96]

While the 1920 Cairo plan was practically shelved, the Tanzim Department continued looking for ways to address congestion. The department proposed widening al-Khalij Street and opening two major arteries, with the goal of alleviating pressure on al-Muski Street—the busiest commercial thoroughfare in the old city—and 'Abbasiyya Street—the city's main link to the northeastern suburbs. These projects were implemented in partnership with the Tramway Company, which was slated as one of the main beneficiaries, since the government planned to authorize the extension of tram lines along the new streets.[97] Financial negotiations between the government and the private monopoly had been ongoing since World War I on a range of issues, including the rising costs of materials and fuel, ticket prices, and the tramway workers' strike of 1919, which paralyzed the city's transportation during the revolution.[98] In 1924, the two parties agreed to use a portion of the company's profits toward road construction.[99] The new streets, named al-Azhar and al-Amir Faruq, were inaugurated in 1929.

As the Egyptianization of the bureaucracy progressed rapidly, the situation became increasingly uncomfortable for the British engineers remaining in government service. One such engineer was Mann Alcock, who worked at the Tanzim Department. In January 1925, Alcock wrote a confidential memorandum criticizing the Tanzim's performance since coming under Egyptian leadership. He lamented the indefinite postponement of the 1920 urban planning scheme, arguing that Cairo had become a "heterogeneous grouping of semi-isolated quarters" rather than a well-connected city.[100] Alcock was concerned about the increasing tide of popular antipathy toward the foreign monopolies controlling tramways, water, gas, and electricity, describing it as a symptom of "xenophobia" coupled with an "ingrained belief in the vampire nature of foreign capital."[101] For him, what added insult to injury was the "wholesale and fanatical adoption of Arabic as the sole medium of official correspondence, even in dealing with matters of a highly technical nature."[102] The memorandum read like a message from the last imperial outpost in the Public Works Ministry. The days of British control over government ministries were numbered.

One of the most important urban planners of the post-revolutionary period was the Egyptian civil engineer Mahmud Sabri Mahbub. Educated in both Egypt and Britain, he worked for the Drainage Department in Leeds, England, and later joined that city's town planning office. In 1919, he returned to Egypt to join the Tanzim Department, eventually becoming head of the department's Town Planning Service in 1934.[103] In this capacity, Mahbub prepared a planning scheme for Cairo in which he proposed a number of far-reaching measures, including the demolition of the entire working-class neighborhood of Bulaq and the relocation of its residents.[104] In his vision, Bulaq would be transformed into a neighborhood for the "better class," featuring commercial street-fronts, multi-story residential buildings, and a number of public parks.[105] In this sense, Mahbub's plan was a continuation of the kind of planning by engineers that Geddes had criticized in India.

Conclusion

This chapter has explored the relationship between infrastructural engineering and urban planning by showing how the 1920 Cairo plan emerged as a process of scaling up practices begun in tandem with the construction of the city's drainage network. Sewering Cairo required close collaboration between sanitary and road engineers on a wide range of issues, including congestion, traffic, street layout, and land use. Ultimately, this collaboration fostered an institutional framework for viewing urban space as an interconnected whole, thereby paving the way for more comprehensive government interventions. However, as the work of Geddes shows, this vision of planning as an outgrowth of engineering was limited by its top-down approach to urban problems and its single-minded focus on slum demolition, road construction, and urban growth.

The discussions surrounding Cairo's sewer system underscore the centrality of urban drainage to the water consumption revolution outlined in this book's introduction. Critics of the project argued that it underestimated the degree to which patterns of water consumption would change in the future, warning that the system must be sufficiently large to accommodate growth. As some critics pointed out, it was untenable to believe that most Egyptians would continue to use only forty-five liters of water daily or rely regularly on mosque latrines and public baths out of necessity. Instead, they argued that the system would drive an increase in domestic water consumption. In this way, these criticisms echoed the debate between Baldwin Latham and government engineers more than two decades earlier (see Chapter 2). As the next chapter will show, the Cairo Water Company was shrewdly cognizant of these issues. Anticipating the imminent rise in demand for its services, the company changed its subscription system to capitalize on this opportunity. In fact, Cairo's drainage network was the single most important economic stimulus for the private water monopoly.

SIX

Pricing the Nile

Meters and the Political Economy of Water

The legendary collaboration between the poet Badiʿ Khayri and the singer Sayyid Darwish produced several songs that memorialized the events of the 1919 revolution from a popular perspective.[1] One of these is the "The Striking Workers' Melody," which invoked the labor strikes that paralyzed the Egyptian capital during that year.[2] Two lines of this song are especially germane to a hydrological history of Cairo:

> *You look at us and see a faucet, you look at him and see a drain*
> *We live in the Nile Valley, but drink by the meter, millimeter by centimeter*[3]

Workers were like faucets in their ability to constantly produce value, while business owners were like drains, siphoning away that wealth into financial reservoirs. As the sewage system began to operate in 1915, the drain became a suitable metaphor for capitalist exploitation. The second line targeted water meters, an emblem of the despised Cairo Water Company and a daily reminder that access to Nile water was controlled by a foreign monopoly. Meters were a typical feature of the company's standpipes, where filtered water was sold to the public at the rate of twenty-seven millimes per cubic meter—the maximum price authorized by the Egyptian government. Sensing the pulse of the street, the song vented frustration over measuring water down to millimeters and centimeters. This strict measurement created a sense of scarcity starkly at odds with the abundance of the Nile, from which the company sourced water at no cost.

In 1919, there was talk that the company might expand the use of meters, potentially mandating their use for all customers. In fact, this is what transpired the following year. Previously, customers wishing to connect their homes to the service had a choice between two subscription methods: a flat monthly rate or billing based on meter readings. From 1920 onward, however, all new customers were required to install water meters and were billed strictly according to their consumption. The company justified this policy change as a measure to curb waste, particularly in light of the recently completed sewage system. In the past, the company argued, customers were constrained by a physical limit—the size of their buildings' cesspools—which restricted the amount of water that could be used. Now that the sewers had removed this constraint, there was no longer any cap on consumption. The company warned that this could lead to careless overuse, so it introduced a financial limit to replace the old physical one. Known as "volumetric pricing"—charging for utilities such as water, gas, or electricity based on the amount used—this strategy was purported to create an incentive for conservation. By making customers pay exactly for what they consumed, the meter would serve as a constant reminder to use water more judiciously.

The water meter quietly irritated the public for years, until the 1930s, when it triggered the most turbulent era for the water company. In 1932, the owners of an apartment building in Fajjala and the nearby Coptic Hospital filed a joint lawsuit against both the company and the Egyptian government, in its capacity as the grantor of the concession. They argued that the new policy violated their "right" to choose between meters and monthly flat rates. In the years that followed, the company faced similar lawsuits, with the government also implicated. In 1934, the company was denounced in the Egyptian Parliament by ʿAbd al-ʿAziz Nazmi, the long-serving representative of Sayyida Zaynab. Proclaiming it to be "the most dangerous monopoly company to the public," Nazmi exposed a wealth of information about the company's history, profits, and the lawsuits it recently faced.[4] This new information was widely picked up by the press, further fueling public outrage. Seeking to contain the situation, the government appointed an expert to arbitrate the dispute concerning water meters, with the choice falling on Sidney Raffety, a British engineer with experience in both the technical and economic aspects of water supply. After investigating the matter, Raffety produced a comprehensive report in which he condemned the company's profit-driven approach, particularly given the vital nature of its service to public health, and recommended capping the profits at a "reasonable" level.

This chapter explores these events, situating them within the broader context of Egypt's changing political economy. As will be seen, this campaign was part of a larger effort to limit the power of foreign corporate monopolies in Egypt, many of them in control of essential services and national resources—the Suez Canal Company being the most infamous example. Although the government initially disregarded Raffety's recommendations, his findings played a significant role in shaping subsequent efforts to regulate all public service monopolies in Egypt. Thus, the campaign against the water company helped shape the antimonopolist mood of the 1940s and laid the groundwork for the broader rejection of the old concessionary system on the eve of 1952, when the Free Officers seized power.

Raffety's arbitration offers a case study for advancing the conversation between engineering and economics. As an expert, his mission was to evaluate how water was priced in Cairo, and therefore, to help the government determine a price that would be fair to both the company and the public. As previously argued, economic analysis is an essential component of what engineers do, so there was nothing inherently exceptional about Raffety's mission in Egypt. What is eye-opening, however, is how he disassembled, then reassembled, the relationship between the company's subscription policy and the construction of the sewage system. As he explained, the switch to volumetric pricing had little to do with conserving water. Despite the intuitive appeal of the conservation argument, this was a shrewd business strategy, calculated to augment profits. Whereas Chapter 2 highlighted the potential rise in demand for water services after the sewers were built, this chapter examines how the water monopoly exploited this demand for maximum gain. Echoing Thorstein Veblen's *The Engineers and the Price System*, this case study of engineering-economic analysis in Cairo underscores the predatory behavior of business enterprise.[5]

Policy Changes

The Cairo Water Company's subscription system remained unchanged from 1887 to 1920. Under the 1887 agreement, customers were given the choice between a monthly flat rate or metered supply (Chapter 1).[6] Flat rates varied based on the type of building and the number of occupants, with the baseline set at thirty-two piasters per month for a household of five, with an additional six piasters for each extra person. Buildings with multiple housing units received a 20 percent discount. Lower rates applied to non-residential buildings such as schools, hospitals, mosques, fountains, and public baths. The key feature of this billing method was its unlimited nature—the monthly rate was fixed regardless of water consumption, allowing customers to use as much water as they wanted. The only real constraint was the size of the building's cesspool. Ignoring this limit, customers would have to deal with the unpleasant situation of sewage overflowing in their basement.

The second subscription method was metered supply, which differed from the flat rate system by billing customers based on actual water usage as recorded by a meter. Customers who chose this option were required to install meters at their own expense. The rate was set at twenty-seven millimes per cubic meter, the maximum price agreed upon by the government and the company. Metered supply was less popular among the company's subscribers, as it was perceived to be more expensive. However, it was preferred in special circumstances, such as seasonal occupancy, when it would not make much sense to pay bills for the whole year. Owners of a winter house, for instance, would pay only for the few months spanning their residence, paying nothing during the rest of the year. As a technology, water meters were expensive, bulky, and unreliable up until the turn of the twentieth century, when they were improved and calibrated for residential use.[7]

The water company began modifying customer contracts in 1898, when William Willcocks, then acting as the company's manager, drafted a proposal for a sewage system in Cairo (Chapter 4). At that time, the company introduced a new clause stating that "if a sewage system is built, the company reserves the right to require the customer to install a water storage cistern at their own expense, to prevent water from being wasted."[8] The "cistern" is that part of the toilet where water is stored for flushing. While today it is more commonly attached to the toilet, in older models it used to be mounted high on the wall and operated with a pull chain. By storing several liters of water, cisterns created the momentum necessary to flush the toilet effectively. In 1871, British authorities found many cisterns to be defective, causing water to leak into the bowl and down the drain. During the following decade, the cistern was improved with a siphon mechanism that solved the issue. This new type of cistern was called a "water waste preventer."[9]

Assuming good intentions, the new clause in customer contracts identified a significant issue but missed its mark. Had the sewage system been constructed, customers would have faced a new reality—they could use as much water as they wanted without worrying about wastewater overflowing in their basements. However, focusing solely on the toilet was misguided. In principle, the same concern about waste applied to other areas of household water consumption, whether in meal preparation, handwashing, or bathing. Moreover, a flushing cistern was already part of the government's minimum sanitary requirements for sewer connection, should such a system be implemented in the future.

During the sewage system's construction, the company once again revised the fine print of its contracts. This time, it reserved the right to require customers to install either a flushing cistern or a water meter. Additionally, a new clause was introduced, allowing the company to increase monthly fees if "it became clear that the value of water used exceeded the value of the monthly subscription."[10] By 1915, the sewage system was completed, but it was not until 1920 that it began operating at full capacity. By then, government properties, commercial buildings, and houses in affluent districts were connected to public sewers.[11] In that same year, the company made one final revision to its contracts, now reserving the right to mandate water meter installation if "the house subject to this contract was, or will be, connected to the sewage system."[12] The second clause, allowing fee increases, remained unchanged. With these adjustments, the company had effectively overhauled its subscription policy. From now on, all new customers had to install water meters. Due to a general lack of public understanding of the company's legal rights and obligations, new customers signed their contracts under the belief that the company was entitled to enforce these terms. As it turned out, however, the legality of these claims was, at best, questionable.

The company justified the new policy on the grounds of water conservation. Before the sewage system, the size of cesspools imposed a physical limit on how much water could be used privately. With the sewers in place, however, consumers faced a different set of incentives. Theoretically, it was now possible to leave the water running all day without the risk of flooding the basement. The company argued that the water meter would serve as a corrective measure, replacing the old physical constraint with a financial one. By charging customers according to their consumption, the meter would serve as a constant reminder to turn off taps, avoid unnecessarily long baths, and repair leaking fixtures. However, as later developments would show, these seemingly reasonable arguments came under scrutiny. Critics revealed that the company had effectively replaced the existing billing system with another that was significantly more expensive.

The shift to metered supply was neither immediate nor retroactive. The new policy applied only to customers who signed up after 1920. Existing customers with flat-rate contracts could not be forced to install meters, and their contracts automatically renewed every few years. Nonetheless, the policy affected existing customers in various situations: if they moved to a new house, if their house was demolished and rebuilt, or if a customer passed away and their descendants wanted to transfer the contract. In these cases, the installation of a meter was mandatory.[13] As a result, the number of customers on metered supply steadily grew from 1920 onwards, while those on flat-rate contracts gradually declined. Although older customers held on to their flat-rate subscriptions for as long as possible, the transition to metered supply played out over the course of the 1920s and 1930s. By 1936, 45 percent of buildings connected to water supply were fitted with meters.[14]

The Age of Private Water Consumption

By changing its policy in 1920, the Cairo Water Company positioned itself to benefit from the anticipated rise in demand for water services. As previously discussed (Chapters 2 and 5), the question of future demand, or future consumption, was initially considered a matter of educated guessing if raised at all. As Baldwin Latham and others argued, the introduction of drainage would transform the city's hydrology, normalizing domestic water usage and leading to increased consumption on average. By the early decades of the twentieth century, there was no doubt among the experts that sewage systems triggered higher demand for water services. Developments after the system's implementation in 1915 confirmed this point of view. By the 1920s, Cairo was entering the era of private water consumption, with a growing number of buildings becoming connected to both water and sewers. In fact, the sewage system was so well-received that it required expansion much sooner than anticipated, a possibility that some engineers had warned of. While the system's designers were cautious about the feasibility of widespread private water consumption, the system's success compelled the government to double its capacity within two decades. Before the system's construction, only 15 percent of buildings were connected to piped water. By the late 1940s, nearly all of Cairo was connected to water and sewers.

When the government built the system in 1915, it sewered the city's main streets, leaving secondary streets to be addressed in the future on a need-to-do basis. From then until the 1940s, all streets were gradually sewered. Each year, the government awarded contracts to local firms to extend public sewers into previously unsewered areas.[15] As early as 1921, the Drainage Department was "being constantly asked by every class of the community to fix more gullies in all areas of the city."[16] By 1929, the government had doubled the budget for laying new sewers. However, officials continued to receive complaints from property owners who wanted to connect their homes but lacked sewers in their vicinity.[17] By 1933, the total length of streets in Cairo had reached 1,000 kilometers, with 650 kilometers supplied with sewers.[18] According to the government's almanac for that year, these developments signaled a shift in public perception: "Property owners have realized the benefits of connecting their premises to the public sewers and the abolition of the old system of cesspit drainage."[19]

The same conclusion was evident from the growing number of applications for sewer connection. In 1921, there were 621 applications; by 1926, this number had increased to 1,376, and by 1931, it had reached 3,931.[20] A House Connection staff member reviewed each application, approving it only after inspecting the building to ensure it was connected to the water supply and met hygienic standards.[21] Additionally, the government considered a house drainage law that would require property owners to connect their premises to public sewers if they were within a certain distance. However, the Drainage Department believed that most property owners would voluntarily connect their houses, given the favorable reception of the system.[22]

These developments had considerable implications for domestic water consumption. In 1906, the Cairo Water Company supplied the city with nineteen million cubic meters of water; by 1928, that figure had risen to thirty-six million cubic meters. This increase outpaced the rise in population, which grew from approximately 650,000 to one million during the same period. Together with the building industry, the sewage system enabled greater water usage per capita. As government officials noted:

> The increased use of water is due, firstly, to the facilities offered by the drainage system for getting rid of water and other waste products, and, secondly, to the greater prosperity of the country which is to be observed by comparing the class of poor properties, which abounded in the city in 1906, with the modern first class buildings that are now being constructed and fitted with up to date sanitation.[23]

In 1939, Mahmud Wasfi, deputy head of the Drainage Department, stated that Cairo was consuming an average of ninety liters of water per capita per day—double the amount estimated in 1907—and that it was expected to consume even more in the future.[24]

This realization was the key factor in the government's decision to expand the capacity of the sewage system, alongside three other contributing factors. First, after an exceptionally high Nile flood in 1916, the government decided to drain the floodwater using the sewage system, which reportedly placed it under considerable strain.[25] Second, the government noticed that the cement of the main collector began to corrode and required extensive repairs.[26] Finally, Cairo's population increased more than anticipated. The original plan was designed for a population of nearly one million by 1932, but Cairo reached this benchmark in 1927. As a result, the government significantly expanded the sewage system to accommodate the additional demand. The new drainage works included a second pumping station, a duplicate main collector, a duplicate rising main, and an emergency storage reservoir. Completed in 1929, these facilities cost the government over half a million Egyptian pounds.[27]

The Campaign Against the Water Company

Customers of the Cairo Water Company typically signed their contracts without scrutinizing the details, as there were no independent means to verify the company's claims. However, a lawsuit was filed in 1932 that changed this dynamic. The case was brought by two plaintiffs: the Queen Nazli Real Estate Company—an association of Egyptian and European owners of a multi-story apartment building in Fajjala—and the nearby Coptic Hospital, which had been built in 1926.[28] Located at 127 Queen Nazli Street (now Ramsis Street), the building contained fifty-one apartments and sixteen stores. Shortly after construction was completed in 1931, the building owners contacted the water company, requesting to have the building connected to the water supply. It appears that the owners were aware that some of the company's old customers were still on monthly flat rate plans, as they specifically requested this type of subscription. However, the company stated that metered supply was the only option available. Despite the owners' objections, the company maintained that it had the right to set rates based solely on meters.

By October of that same year, the building was already connected to the water mains, but negotiations with the water company were still unresolved. According to the lawsuit petition, the company threatened to cut off the water supply if an agreement was not reached. Under pressure, the building owners reluctantly agreed to metered supply on an experimental basis for one year, with the understanding that the rate would be adjusted thereafter. A year later, the owners realized that they were paying significantly more than they would have under the old flat-rate system—forty-seven Egyptian pounds annually compared to an estimated thirteen. When they protested again, the company assured them that the rate could be adjusted based on "the average price during the experimental year."[29]

This time, however, the owners would not back down. Alongside the Coptic Hospital, they decided to take legal action, arguing that the Cairo Water Company had violated their right to choose between monthly flat rates and metered supply. They also sued the government, holding it accountable as the grantor of the company's concession. The case was referred to the Mixed Courts on December 19, 1932. Over the following year, as the case progressed, K. Boulad, the plaintiffs' lawyer, reportedly uncovered substantial evidence against the company. For some, this explained why the company settled the case just before a ruling was expected, while generously compensating the plaintiffs.[30]

The company's attempt to hide the evidence and avoid public exposure failed, as the issue quickly surfaced in the Egyptian Parliament. Somehow, 'Abd al-'Aziz Nazmi, the representative for Sayyida Zaynab, was able to obtain the information gathered by Boulad. Nazmi was a doctor by training and a prominent public figure at the time. Educated in France, he returned to Egypt to work in the administration of charitable endowments, eventually rising to the position of chief doctor. He was also a prolific writer on pediatrics and actively involved in several grassroots, educational, and youth committees.[31] After receiving the lawsuit information, Nazmi conducted further research. This led him to the library of the Egyptian Senate, where he found a rare volume titled *Recueil Général des Contrats* (*General Collection of Contracts*), a printed book containing all government contracts.[32]

In a fiery speech before Parliament on April 24, 1934, Nazmi accused the Cairo Water Company of ripping off the public while freely exploiting the water of the Nile. Armed with a wealth of evidence, Nazmi said he was ready to substantiate these claims. Over that session and the next, Nazmi walked the Parliament through a series of disastrous agreements of which the public knew little or nothing. He began with the original 1865 concession, revealing Nubar Pasha's dual role as both the government's representative and one of the water company's major shareholders.

strengthened the company's position. Following the ruling, the Public Works Ministry decided to bring in a foreign expert to arbitrate the case. The expert required was an engineer with experience in both the technical and economic aspects of water supply in cities of comparable size. In 1937, British engineer Sidney Raffety was selected.

Raffety was a specialist with a wide range of experience in waterworks. In his early career, he worked for both municipal bodies and private water companies in various English towns. During the First World War, he served with the Royal Engineers in Macedonia and Turkey. Upon his return, he steadily rose to prominence as a consultant, fostering ties within the British Parliament. His obituary notes that "from 1939 to 1945 there were few water bills with which he was not concerned."[51] Beyond his work in Cairo, Raffety's other major foreign consultancy took place in Palestine (in 1963–65), where he designed waterworks for Jerusalem, surveyed the water resources of Nablus, and analyzed the hydrology of the West Bank.

Expert Arbitration

The government hired Raffety to arbitrate between the company and the public, specifically asking him to address two interrelated issues of economic significance. First, they requested an estimate of "the cost of producing one cubic meter of filtered water." This information would serve as a solid reference point for evaluating the current price of water—twenty-seven millimes per cubic meter. Second, Raffety was asked to offer an opinion on what would constitute a "reasonable margin of profit" for the water company.[52] In essence, the government sought his expertise to distinguish between legitimate and excessive profits, and to provide a comparative analysis of profit margins for water companies in similar cities. Notably, Raffety calculated the cost of producing one cubic meter of filtered water, revealing that it averaged a mere 3.32 millimes between 1932 and 1934. This significant gap between the cost of production and sales price was what generated huge profits, ultimately ending up in the pockets of the company's shareholders.

The idea that the price of goods and services should be judged against the cost of their production has long been common sense. In classical political economy, the price of commodities reflected—however imperfectly—the cost of various production inputs, including materials, land, equipment, and labor.[53] Pricing a product, in this view, was a mechanism for recovering what the supplier had already spent to put a product on the market, along with some return on capital investment.[54] The reason an experienced engineer was needed in this case was that calculating the production cost of water was a complex task requiring deep familiarity not only with how the distribution network worked as a whole, but also with its individual components—pumps, pipes, filters, labor, and consumable supplies—each of which had a different lifecycle in the production process.

In fact, this way of analyzing water pricing was characteristic of an approach that many other engineers around the world adopted when considering the pricing of other goods and services.[55] As the scale of industrial systems grew larger by the beginning of the twentieth century, the engineers managing these systems often experienced a contradiction between what business executives asked of them and what made sense from a technological standpoint. In the interest of cutting costs and maximizing profits, business managers often sought to disrupt the market by creating various disturbances and artificial limits on the supply of products and services. For a growing number of engineers, however, these limits had little to do with the actual capacity of industrial systems to produce goods and services. In effect, the engineers argued that business executives were in the habit of introducing inefficiencies designed to create profits, regardless of what was technologically possible. In *The Engineers and the Price System*, Veblen dwells on the gap identified by engineers between society's technological resources and the prevailing direction of the economy under business management. For him, it showed that collective resources were being "sabotaged" in the interest of maximizing private gains, further confirming his view on the predatory character of business enterprise under the conditions of modern capitalism.[56]

In Cairo, Raffety exposed how the Cairo Water Company exploited the construction of the sewage system as a business opportunity to disrupt the market. The disturbance or artificial limit introduced in this case was changing the subscription system, which forced new customers to pay by the meter. Just as the American engineers studied by Veblen pointed out the gap between technological capabilities and the way things were actually run by business executives, Raffety pointed out the gap between the cost of producing filtered water in Cairo and the way it was priced by the private company. In other words, he exposed how the water company exploited the public for profits without any regard for the actual expenses of supplying water to them—a discovery all the more glaring given the vital importance of water supply for public health.

Raffety visited Egypt from November 10 to November 28, 1937. During this time, he met with several officials from the Public Works Ministry and the Cairo Water Company, including Sadiq Henein, the company's director. He was granted access to inspect the company's facilities and other waterworks across the country. In addition, he obtained copies of the agreements between the government and the company, the company's printed accounts dating back to 1870, and detailed records covering the period between 1923 and 1936.[57] But when he requested figures showing the production cost per unit of filtered water in Cairo, company officials claimed that they did not have this information themselves.

In the past, the company had been reluctant to provide the government with detailed information regarding its resources and expenses. With Raffety, they claimed their accounts did not differentiate between filtered, unfiltered, and underground water, nor did they differentiate between Cairo and the town of Asyut—where the company had acquired the right to distribute water in 1907. In his report, Raffety admitted that calculating the cost of each class of water in Cairo could be challenging, but he argued that a "close approximation" was nonetheless feasible.[58] Using the data he had collected, he prepared three extensive tables detailing annual receipts, expenses, and quantities of water supplied. From these tables, he was able to isolate filtered water and calculate both the production cost and profit per cubic meter (Table 2). Based on this, he drew several conclusions. First and foremost, there was a large difference between the production cost of filtered water and the price charged by the company to the public. Second, the cost per unit had generally decreased since 1923. Finally, the profit per unit had steadily increased over the same period.

TABLE 2. Cost and Profit Per Cubic Meter of Filtered Water[a]

Year	Cost per cubic meter (in millimes)	Profit per cubic meter (in millimes)	Ratio (per cent) of expenses to receipts
1923	4.61	5.89	42.45
1924	4.07	6.29	39.34
1925	4.10	6.51	38.67
1926	-	-	-
1927	3.65	7.19	33.65
1928	3.70	7.20	34.00
1929	3.54	7.55	31.91
1930	3.68	7.95	31.65
1931	-	-	-
1932	3.40	8.07	29.64
1933	3.50	8.12	30.14
1934	3.08	8.24	27.05
1935	-	-	-
1936	-	-	-
Average for 1932–34	3.32	8.14	

[a]The table shows columns 1, 9, 10, and 11 of one of the larger tables prepared by Raffety.

Source: DWQ, Majlis al-Wuzara' (Council of Ministers), file no. 0075-059295, "Report to his Excellency the Minister of Public Works on the Charges for Water in Cairo," by S. R. Raffety, February 4 (1938).

Raffety considered the water company's profits in light of its financial structure, arriving at a startling conclusion. He estimated that current shareholders were earning 70 percent in dividends, and projected that, assuming a continued increase in water supply, dividends would reach 90 percent by 1941.[59] To provide context, Raffety referenced the financial regulations governing private water companies in England. The Waterworks Clauses Act of 1847 capped the dividends of all English water companies at 10 percent, with any additional shares issued after a company's formation receiving even lower dividends, typically 7 percent. According to the same act, if a water company's revenues exceeded the amount needed to pay maximum dividends and contribute to reserve funds, it was legally required to reduce its charges to the public.[60] While Raffety acknowledged that "English standards" could not be applied universally without considering political and economic contexts, the comparison highlighted an essential principle: a private company holding a monopoly over a public utility like water should not be permitted to make excessive profits.[61]

Raffety analyzed the Cairo Water Company's profitability since its establishment, dividing its history into two main periods: 1) from 1888 to 1920, during which the company's revenues steadily increased, and 2) from 1920 to 1936, marked by a more rapid rise in revenues.[62] He explained why the turning point took place in 1920:

> It is significant that since 1920, when the establishment of main drainage in Cairo became imminent, the Water Company has adopted a policy of introducing metered supplies in substitution of unmetered supplies charged on the basis of the number of occupants. It is obvious that not only the introduction of main drainage but also the improvement of housing conditions generally would increase the average quantity of water used per person, which is a feature not peculiar to Cairo. The older method of charging must therefore have been expected to result in a reduced revenue per cubic meter.[63]

By changing its policy in 1920, the company aimed to preempt a potential decline in profit margins and capitalize on the anticipated growth in water usage. Raffety also confirmed that profits from metered supply were "greater for the same quantity of water used" compared to flat-rate subscriptions. While the new policy was alleged to compensate for the anticipated decline in revenue per unit, it ultimately propelled the water company into an era of even greater profitability.

Raffety was not persuaded by the water company's argument regarding waste. While he acknowledged there was some validity to the claim, he presented several other important factors to consider. First, waste "can also be kept down almost equally well by careful inspection."[64] Second, meters tend to reduce consumption levels only in the short term. Upon initially converting from unlimited to metered supply, customers might avoid using too much water. Over time, however, they become less concerned and return to previous habits, unless there are significant leaks from their pipes and fittings. Finally, if water prices were set too high, particularly for small consumers, it would undermine the goal of sanitation. As Raffety pointed out, "ordinary house supplies in England are not permitted to be given by meter."[65] Ultimately, Raffety doubted that "the gradual introduction of meters will have much effect on the quantity of water used."[66] Instead, he anticipated that water consumption would continue to rise, as it had in recent years, especially if the number of buildings equipped with modern conveniences continued to grow.

Toward the end of his extensive report, Raffety turned his attention to the recent negotiations between the government and the Cairo Water Company. He noted that most customers still relied on the old subscription method: in 1936, there were 18,351 metered establishments and 22,143 on flat rates. Since metering was more expensive, Raffety argued that "if and when all should be supplied by meter the result would be a large increase in the revenue of the water company."[67] But what about universalizing the meter while reducing the rate to sixteen millimes? This was where negotiations between the government and the company had stalled. Raffety demonstrated that this agreement would generate greater profits for the company than it was currently earning. In fact, universalizing the meter at a rate of 13.5 millimes would still increase revenues, and at twelve millimes, the company would maintain its current level of profit. Ultimately, Raffety concluded that "so long as Cairo continues to grow, the receipts and profits of the company must increase if the price for water is not reduced from time to time."[68]

For Raffety, there was no justification to increase the company's already "excessive" profits. In fact, he argued, there was "much to be said for their reduction."[69] The solution, in his view, was not simply to lower the water rate, but to cap the profits the company was permitted to earn and distribute. By doing so, the government could ensure a continuous reduction in water prices and avoid disagreements over what a fair rate might be.[70] Raffety outlined several practical steps should the government choose this approach, including a schedule for reducing the company's profits over the coming years.

In February 1938, Raffety submitted his report to the Public Works Ministry, which then conveyed its findings to the Cairo Water Company. In April of the same year, the company replied, categorically rejecting the principle of "reasonable profits" as the foundation for a new agreement with the government.[71] The company argued that nothing in the original concession or subsequent agreements supported this principle, and that it directly contradicted its right to exploitation. Furthermore, the company maintained that it would be easier to agree on a new water rate than to resolve opinions about what constituted reasonable profits.

On July 4, 1938, and after another round of negotiations, the government and the water company signed a new agreement with the following terms: 1) the price of water was reduced to 14.5 millimes for private consumers; 2) within two years, flat rate plans would be discontinued; 3) starting in January 1940, all profits exceeding L.E. 400,000 would be divided 60/40 percent between the company and the government, with the government's share used to further lower water prices; 4) the company would make water meters available for rent at a price that yielded no profit to the company; 5) the government would have the authority to oversee customer contracts; and 6) the geographical boundaries of the concession would be clearly defined, as "Cairo and its suburbs" continued to expand in new directions.[72] These terms clearly disregarded Raffety's recommendations, which explains why the government never published his report and withheld it from Parliament for several years. The government defended the agreement as a compromise, especially considering the initial 1936 Mixed Courts ruling. With the appeal still pending, the government feared that its negotiating position would be even weaker if the final ruling favored the company.[73]

Parliamentary Challenges

Barely ten days had passed since the new water agreement was signed when it led to a standoff in the Egyptian Parliament. Several representatives criticized the government for signing the agreement, and called on their colleagues to vote against it. Although they failed to carry a motion, they succeeded in raising questions about the agreement's constitutionality. The conflict pitted members of the Wafd Party—the most popular since the 1919 revolution—against an ex-Wafdist who had recently defected to one of the so-called "minority parties"—small factions that sometimes united to offset the Wafd. As is well known among historians of Egypt, the thirty years following the country's declaration as a constitutional monarchy in 1922 were marked by constant maneuvering among the King, the British, and political parties, each vying for control over the executive branch.[74] In July 1938, the government was formed from several small parties that allied themselves with the newly enthroned King Faruq.

On one side of the standoff was the ex-Wafdist Ahmad Mahir, in his capacity as the current Minister of Finance. Mahir held degrees in law and economics and was a former Minister of Education and chairman of the Parliament's Budget and Accounting Committee.[75] On the other side of the confrontation were four prominent Wafdists: Mahmud Abu al-Fath, a journalist known for his activism in 1919, founder of the party's mouthpiece *al-Masri,* and future head of the journalists syndicate (in 1941); 'Abd al-Hamid 'Abd al-Haqq, who later became head of the lawyers syndicate and minister in several governments; Mahmud Ghannam, a lawyer and future Minister of Commerce; and Yusuf al-Jundi, a lawyer and former "president" of the self-declared "Zifta Republic." During the 1919 revolution, the provincial town of Zifta briefly declared independence from British-controlled Egypt, only to be retaken by an Australian fleet from the colonial army.[76]

Abu al-Fath criticized the government for not upholding the public's right to choose between flat rates and meters.[77] He argued that article 6 of the 1887 agreement stipulated that the company must offer this option to customers. Additional evidence supported this claim, including the fact that the company had maintained this policy until 1920, the legal opinion of 'Abd al-Hamid Badawi from Government Litigation, and the generous settlement with the Queen Nazli Building and the Coptic Hospital. Abu al-Fath proposed alternatives to universal metering at a single rate, such as installing meters in wealthy districts while allowing poor ones to continue on flat-rate plans, or introducing progressive pricing based on the property's rental value—methods successfully implemented in other countries. While acknowledging the company's victory in the Mixed Courts, he remained confident in the appeal.

In response, Mahir turned to legal textualism, going back to the original concession of 1865. He pointed out that the concession used the word "pourra" rather than "devra," explaining that anyone familiar with

company gained more by canceling monthly contracts and universalizing meters. Third, citing Raffety's report, the committee pointed out that the company would still be profitable if the meter was universalized at a price of twelve millimes.[89] Despite these findings, the government opted to wait for the resolution of all lawsuits against the company before taking further action.

In June 1941, the Mixed Court of Appeal overturned the initial ruling in favor of the Cairo Water Company, affirming that subscribers had the right to choose between flat rates and meters.[90] While this decision represented a victory for the public, it was largely a moral one, as meters had already been installed citywide over the past few years, in line with the 1938 agreement. In practical and logistical terms, metering was now irreversible. The government entered a new round of negotiations with the company, aiming to establish a new agreement to be ratified by Parliament. In 1946, Parliament passed the new agreement, which included the following terms: 1) the price of water was reduced from 14.5 to ten millimes; 2) the company and the government would continue to share profits at a 60/40 percent ratio, but now applying to profits exceeding L.E. 350,000, and with the government's share allocated to the same public fund established earlier.[91] The agreement was signed by Public Works Minister ʿAbd al-Qawi Ahmad and the company's chairman, ʿAli al-Shamsi. Notably, although the new deal successfully reduced the price of water, it still did not place limits on the company's profits.

The Road to Nationalization

The Cairo Water Company was liquidated in July 1957, and its assets were seized by the Egyptian state—part of the sweeping transfer of ownership from private to state control following the nationalization of the Suez Canal. On July 23, 1952, King Faruq had been ousted by a coup led by a group of military officers. According to the conventional narrative of Egyptian historiography, only the military's involvement in politics was able to end colonial presence in Egypt, with British forces finally being expelled from the Suez Canal Zone. In this view, the nationalization of the Suez Canal and other foreign monopolies epitomized Egypt's eventual triumph over Western imperialism and its economic tentacles.

However, this narrative overlooks the history leading up to the coup, downplaying significant continuities with the monarchical period. As seen above, foreign monopolies had faced mounting challenges since the 1930s, coming to represent archetypal symbols of colonial extraction and exploitation. Coupled with the widespread social unrest and mass political action in the 1940s, the position of these concessionary monopolies became increasingly untenable as they faced a wave of government regulations, if not outright calls for nationalization. Thus, the pre-1952 period had already set the scene for the favorable reception of Nasser's nationalization program. In fact, the campaign against the Cairo Water Company and the efforts to bring it under government regulation were integral to these broader struggles.

World War II brought widespread economic and political developments in the Middle East. Wartime trade disruptions led to shortages in food supplies and a rise in the cost of living. In Egypt, the war stimulated import-substitution industrialization and led to the introduction of new tools of economic intervention. By 1945, industrial output had risen by 40 percent compared to 1939 levels.[92] Simultaneously, wartime management of food supplies set an institutional precedent for state management and regulation of the economy.[93] These changes coincided with the rise of new political forces, both to the left and right of traditional parties—including communists, Islamists, and fascists—who sought greater influence over students and the working class.[94] One of the defining political events of this period was the 1946 strike organized by the National Committee of Workers and Students, where thousands of demonstrators gathered in Tahrir Square, raising various social demands and calling for the expulsion of British troops still stationed in Egypt.[95]

Leftist intellectuals criticized not only the concessionary regime first established by Khedive Isma'il and expanded under British rule, but also the monopolistic and oligopolistic structure of the economy since independence in 1922. One of these intellectuals was the Marxist Ahmad Sadiq Sa'd—my maternal grandfather. Born to a Jewish family that had settled in Egypt in the late nineteenth century, Sadiq Sa'd was an electrical engineer by training. In his youth, he joined antifascist groups and became a founding member of the New Dawn communist organization, which was particularly active among workers in the industrial neighborhood of Shubra al-Khayma.[96] In 1945, he authored a short booklet titled *The Tragedy of Supply*, analyzing the sharp rise in living expenses during the war. He argued that the issue went beyond merchants' greed, the presence of Allied military forces, or the corruption of state officials. Rather, it was the result of several short- and long-term factors: speculation, monopolies, a colonial agrarian economy centered on cotton exports, and the isolation of the state apparatus from the popular classes. As a solution, he called for the immediate nationalization of monopolies with a large concentration of foreign capital—including water supply, gas, electricity, sugar, and tram transportation—as well as the gradual requisitioning of industrial sectors producing essential or strategic commodities.[97] Earlier that year, he had published another booklet on the agrarian question, proposing a radical program of land redistribution.[98]

Responding to these external and internal pressures, the Egyptian government began to experiment with a program of broad economic and social reforms.[99] On July 24, 1947, the government passed the Public Service Concessions Law. First drafted a decade earlier, the law was based on the theories of French legal scholar Gaston Jèze—who had been referenced in Parliament back in 1938, when Mahir mentioned a "new legal theory" under government consideration. Jèze argued that concessions over state resources were not limited to the literal terms of agreements but must remain open to legitimate modifications in the name of public interest.[100] At the same time, the arbitration case with the Cairo Water Company served as a crucial frame of reference during discussions in the Egyptian Senate. Legislators showed particular interest in Raffety's statement that English regulators allowed no more than 10 percent profit for water companies, and in his opinion that water pricing should be reduced through periodic adjustments.[101]

The law limited public service concessions to thirty years (Article 1), empowered the government to periodically reduce service prices (Article 2), and capped profits at no more than 10 percent (Article 3). Article 5 granted the government the right to modify agreements in the name of the public good, while Article 6 stipulated that if unforeseen circumstances led to exorbitant profits, the government could adjust service prices to bring profits down to an "acceptable level." However, the law contained a significant loophole: Article 8 was limited by "the provisions of any agreement issued by a law prior to this one."[102] The Senate insisted on including this clause, despite the government's objections.[103] As a result, the Cairo Water Company was excluded from the law, since its 1946 agreement with the government had already passed into law. Therefore, the company's position remained essentially unchanged by the new legislation.

The government implemented other significant reforms. On August 4, 1947, it enacted the Stock Companies Law, which mandated that Egyptians hold a majority stake (51 percent) in all private ventures.[104] The law also set required quotas for Egyptian employees in the labor force, management, and boards of directors. Notably, the law strengthened the position of Egyptian capitalists—now essential for any business venture—but it left unchanged the overarching monopolistic structure of Egypt's economy.[105] Other important social and administrative reforms of that period include the creation of a Cairo municipality in 1949. Placed under the direction of Mahmud Riyad, a notable architect and urban planner, the municipality expanded social housing initiatives and managed essential public services, including gas and electricity, although water supply and tramway transportation remained under private control.[106]

The conventional historiography of Egypt under Nasser overlooks not only conditions on the eve of 1952 but also the early years following the coup. It is well known that the Free Officers had no clear plans for governance. However, it is less recognized that they initially sought to encourage the private sector and reassure foreign investors, with the full support of the United States.[107] In fact, the new regime quickly rolled back some existing regulations. On July 30, 1952, the 1947 Stock Companies Law was modified to reduce the required proportion of Egyptian capital from 51 to 49 percent, thus allowing foreign investors to regain majority stakes. Historian Anwar 'Abd al-Malik attributes this measure to the advice of 'Abd al-Jalil al-'Imari, the Minister of Finance, and Ahmad Husayn, ambassador to Washington.[108] Pro-big business circles praised the decision as ending "that absolutely unjustified fear of foreign capital," which they claimed had hindered Egypt's economic development.[109] Similarly, the Mining Law was revised to make it more acceptable to foreign oil companies.[110] More broadly, the Free Officers' initially good relations with the United States reflected their hostility toward communism and their harsh crackdown on organized labor.[111]

However, as Washington continued to deny Egypt's request for heavy weaponry, Nasser turned to the Soviet bloc. The 1955 arms deal with Czechoslovakia marked the beginning of the end of Egypt's amicable relations with the United States, with repercussions across various policy areas. This shift played a pivotal role in Egypt's failure to secure funding for the High Dam project. Shortly thereafter, Nasser nationalized the Suez Canal, framing the decision not merely as an act of retribution but as a restoration of the canal to its rightful owners, correcting the injustices committed by the Khedives when they signed the concession a century earlier. Following the canal's seizure, the regime fully embraced a Third World populist agenda with an emphasis on social redistribution and independence from economic imperialism. On July 1, 1957, the Cairo Water Company was liquidated, with its assets and facilities transferred to the Cairo Municipality.[112] The decision marked the beginning of a new chapter in the history of water services in the Egyptian capital, one in which they were managed under a public monopoly.

Conclusion

A central claim of this book is that urban water supply cannot be understood apart from urban drainage. Grasping the relationship between these two infrastructures renders intelligible a wide array of otherwise puzzling developments. As earlier chapters have shown, engineers and administrators initially struggled to conceptualize this relationship, prompting protracted debates over how sewer networks might shape future patterns of water consumption. By the early decades of the twentieth century, however, a broad consensus had formed: the introduction of sewerage would radically reconfigure urban hydrology, both in qualitative and quantitative terms. Sewer systems normalized private water use, displacing older environmental regimes grounded in collective facilities and community-level management, while simultaneously driving up average levels of consumption. These dynamics were clearly visible in Cairo after the completion of the city's drainage infrastructure in 1915. As argued, no one awaited these developments more than the Cairo Water Company. With a monopoly on water supply, it stood to gain the most.

To fully exploit this opening, the company introduced a disturbance designed to disrupt the market. By changing the subscription system in 1920, it forced new customers to install meters and pay according to the maximum rate. As Raffety and others demonstrated, the new policy had little to do with water conservation, despite the company's public justifications. Rather, it was a calculated move to boost revenues. And, in fact, it succeeded in ushering the company into a new phase of profitability. At the same time, however, it also dragged the company into a series of legal and political troubles, including the 1932 joint lawsuit by the Queen Nazli Building and the Coptic Hospital, the parliamentary campaign of the 1930s led by Nazmi and other members, the 1946 agreement with the government, and eventually outright nationalization under Nasser.

CONCLUSION

Privatization Redux

In hindsight, the era when Cairo's water services matched the "infrastructural ideal"[1] of a public monopoly was, in fact, a brief interregnum between two eras of privatization. The first spanned the career of the Cairo Water Company (1865–1957), while the second has stretched from the 1990s to the present, when successive Egyptian governments have taken steps to reprivatize the service. While Anwar Sadat, the President of Egypt from 1970 to 1981, laid the groundwork for water privatization by instituting market-oriented economic liberalization policies—known as the "Infitah"—it was Husni Mubarak, the President from 1981 to 2011, who took the first concrete steps toward privatizing government agencies that controlled water services. The turn may be sealed by a recent decision by President 'Abd al-Fattah al-Sisi, who came to power in 2013 in the aftermath of the January 25 revolution, which may enable a privately owned corporation to supply Nile water to Cairo's inhabitants once again, for the first time since nationalization in 1957.

This decision comes at a time when the country's water security is increasingly uncertain. The Egyptian capital is currently mushrooming, approximately doubling in size with the recent addition of the New

Administrative Capital—under construction since 2015. Today, the greater Cairo metropolitan area is home to over twenty-two million people—roughly 20 percent of Egypt's total population—making it one of the largest megacities in the world. As such, it contributes disproportionately to Egypt's water needs, which are estimated at 114 billion cubic meters. Yet available water resources cover only about sixty billion cubic meters—primarily sourced from the Nile (55.5)—while the rest is largely imported in "virtual" water: that which is used for the cultivation of water-intensive crops such as wheat, maize, soybeans, rice, and sugar, which Egypt purchases from the global market.[2] This gap between needs and resources may widen even further due to the construction of the Renaissance Dam in Ethiopia. Thus, while the capital metropolis is becoming ever more thirsty, the current government has opened the door for the private sector to manage the city's water supply as a profit-making enterprise, defying not only Egypt's own historical experience of water privatization—recounted in this book—but also the lessons of the recent past, as demonstrated by failed neoliberal experiments in water privatization in cities around the world.

The interregnum, during which Cairo's water services operated as a centralized, publicly owned, and expertly managed system (from 1957 to 1991), is especially brief in comparison to the experience of major Western cities, where water companies were heavily regulated, municipalized, or nationalized by the beginning of the twentieth century. Thus, by the 1990s when countries around the world began to privatize water supplies under neoliberal pressures, the notion of public monopoly over essential services was more firmly entrenched in Western democracies than in Egypt, where it was only a relatively recent creation, shaped by Nasser's turn to left-wing, authoritarian economic populism. In Europe, public monopoly over water supply was "splintered" against the backdrop of long-held conventions of liberal democracy.[3] In Egypt and other non-Western countries, by contrast, privatization proceeded in contexts where colonial plunder was only a recent memory and democratic institutions were weak or nonexistent.

During the Nasser era, Cairo's water network largely kept pace with the capital's growing demand. After the company was nationalized in 1957, the city's water supply was reconfigured under the management of a government agency—the Cairo Water Service Administration—which inherited all the physical assets of the private company. Placed under the Cairo Municipality, the administration's remit included setting general policies, developing programs for service extension, approving the budget, and determining service pricing.[4] Meanwhile, the city's drainage system came under increasing stress throughout the 1950s and 1960s, with sewage overflow in the streets becoming a regular sight. Urban authorities attributed the issue to both population growth and deficiencies in the system's original design—especially the fact that the city's entire sewage flow was directed toward a single outlet.[5] In 1964, a permanent committee was formed to address the issue and study a project intended to serve the city for the next fifty years.[6]

During the 1970s, the political economy of water provisioning remained the same despite Sadat's reorientation of the Egyptian economy away from the Soviet bloc and toward the United States.[7] With the support of the International Monetary Fund and the World Bank, Sadat encouraged local and foreign investors, reduced state control, cut government spending and subsidies, and incentivized the growth of the private sector. However, his attempt in 1977 to lift subsidies on basic foodstuffs—particularly bread—triggered a wave of protests unlike any Egypt had experienced since the 1919 revolution, prompting him to quickly reverse course. If anything, this experience made his successor, Mubarak, more cautious when formulating policies that directly affected foodstuffs and essential services during the first decade of his rule.

However, as Mubarak continued to borrow money from international economic agencies in order to maintain fiscal solvency, he initiated a policy of gradual privatization of services such as electricity and water supply. Between 1991 and 2004, the government implemented several measures that ultimately restructured the water and drainage administrations into the Drinking Water and Sanitary Drainage Holding Company, with branches that eventually covered Egypt's twenty-eight governorates.[8] By creating a state-owned "holding company," the regime sought to achieve two interrelated objectives. On the one hand, it took concrete steps to privatize urban water supplies while stopping short of ceding total control to private investors, thus avoiding excessive domestic criticism. At the same time, it could present itself as pursuing economic self-sufficiency to its international creditors, and therefore seek additional loans. The main economic principle behind the holding company was to maintain "cost recovery."

As shown in this book (Chapter 2), the idea of cost recovery has long been an established principle in the economic management of water and transportation infrastructures. Notably, however, it was primarily applied to operational expenses rather than capital investments. For instance, when the Egyptian government raised the house tax from 8.3 to 10 percent in 1910, it sought to recover not the initial expenditure needed to build the sewage system, but the annual costs of maintaining and operating it. In any case, even then it was understood that there were several alternatives for cost recovery, each implying a different economic and political agenda: an indiscriminate blanket tax recovered costs differently than a tax proportionate to the wealth and income of users—measured, for instance, by the value of urban property.

The Mubarak regime practiced cost recovery in ways that left unclear exactly what costs were being recovered from the public. In fact, the government's own studies recommended going beyond the running costs and turning water supply into a profitable enterprise, stressing the need "to make a suitable surplus profit for shareholders in order to encourage citizens to invest in this field."[9] In line with neoliberal theories of governance, the creation of a holding company was framed as a necessary economic and administrative reform to improve the performance of Egypt's bloated, inefficient, and economically unfeasible bureaucracy. As neoliberal theory proclaims, privatization would unleash the spirit of competition, leading to better service, more robust investment in the infrastructure, and even lower prices.

As the water sector was being restructured, Cairo witnessed developments at opposite ends of the social spectrum. On the one hand, there was a rapid growth of informal settlements—the *ashwa'iyyat*. By 2000, over half of the city's residents lived in informal areas, often built on private agricultural land.[10] On the other hand, Cairo was expanding into the desert. Beginning from the mid-1990s, the Mubarak regime encouraged urban development as a means to generate economic growth. During the following decade, vast swaths of desert land were sold at nominal prices, often to businessmen close to the regime.[11] The state offered guarantees for extending infrastructures or, alternatively, left it to private developers in exchange for additional discounts. As the work of David Sims has shown, successive post-1952 governments promoted desert development as a panacea for Egypt's social and economic ills.[12] Desert megaprojects were framed as a solution to problems of congestion, housing and food supply shortages, and insufficient economic development. Yet, as the record shows, these projects largely failed to meet the country's needs, instead benefiting a small group of well-connected businessmen and regime insiders in the construction and development sector.

Notably, areas developed in the 1990s and 2000s introduced a new urban typology to the Egyptian landscape: the gated community.[13] The urbanism of the Fifth Settlement and Shaykh Zayed is defined by a shift toward luxurious gated compounds designed as escapes from Cairo's traditional congestion, disorder, and informal areas. These compounds prioritized privacy and prestige, moving away from high-rise urban blocks toward sprawling low-density layouts. They were characterized by a considerable allocation of land dedicated exclusively to high-end recreational amenities and environmental aesthetics. Residents gained immediate access to private pools, artificial lakes, and meticulously manicured lawns. In premium districts, these communities were often built around golf courses, which served both as recreational hubs and as central landscape features for the surrounding villas. Complementing this type of development was the rise of the modern shopping mall, which emerged as the primary commercial and social anchor for these desert cities. Large-scale retail hubs like Mall of Arabia and Cairo Festival City Mall provided "sanitized" and climate-controlled environments for consumption and leisure, replacing Cairo's street life with private, car-oriented commercial centers.

The water sector's limited privatization under Mubarak did not produce the projected results. Instead of improving the service, it pushed Egyptian cities into a new level of water deprivation. Soon after the holding company was created, the price of water doubled in parts of Cairo, leading to sporadic protests. In a 2005 protest in northern Cairo, "angry residents chased bill collectors down the streets."[14] Over the following few years, dozens of protests took place all over the country. In Bani Suwayf, demonstrators raised banners saying "we are thirsty in the land of the Nile."[15] In fact, Egypt's per capita water share had been steadily declining over the last two decades. Around 1990, the country crossed the water poverty line—set at 1000 cubic meters per capita per year—for the first time. By 2007, it had fallen further into water poverty, with an average water share of 700 cubic meters. Thus, on the eve of the 2011 revolution, the Mubarak regime regularly diverted Egypt's already overstretched water resources to gated communities and seaside resorts, supporting the posh lifestyles of the elite.[16]

These polarizing dynamics have intensified during the past decade. Since 2015, the Egyptian government has launched a project to build the New Administrative Capital, a colossal urban undertaking located roughly forty-five kilometers east of Cairo and designed to function as the country's future seat of power.[17] Spanning approximately 700 square kilometers—a footprint nearly as large as Singapore and roughly equivalent to the size of present-day Cairo—the city is planned to accommodate approximately seven million residents, thereby easing the density of the current capital. With a total project cost estimated at over $58 billion, the New Capital embraces a Dubai-style model of urbanism characterized by spectacular, record-breaking architecture. Its skyline is marked by the "Iconic Tower," the tallest building in Africa, while its spiritual and civic landmarks include one of the world's largest mosques and the Middle East's largest churches, as well as the country's tallest flagpole located in one its central squares.

The New Capital is designed to be Egypt's first "smart city," with water systems integrating facilities for desalination, wastewater recycling, smart meters, and real-time monitoring. As of May 2023, fourteen ministries and government agencies have been relocated there. However, the city remains largely unfinished and sparsely inhabited. Whether it succeeds or becomes another failed "desert dream" it is too early to tell, but critics have reasonably questioned the opportunity costs of such an undertaking. With this price tag, for instance, more than nine new metro lines could have been built.[18] Instead, the government has been demolishing centrally-located informal areas—especially in Bulaq and Old Cairo—on a mass scale, forcibly relocating nearly 2.8 million residents to new areas in the desert.[19] At the same time, Cairo's scarce green spaces were destroyed to widen the streets and build bridges, catering to car traffic at the expense of the well-being of the old city's residents.

Egypt's water security has been in doubt since 2011 due to the construction of the Renaissance Dam in Ethiopia. Finished in 2023, the dam has strained diplomatic relations with Sudan and Egypt, the two downstream countries, with the rhetoric of regional armed conflicts over water resources periodically flaring up during the last decade. Concerned about a potentially critical reduction in its water supply, Egypt continues to assert its right to defend its water share based on the 1929 and 1959 agreements between riparian countries. Ethiopia, on the other hand, insists on its right to develop its own water resources, framing the dam as essential for generating electricity and addressing the country's energy shortages.[20] Since the dam was only very recently filled (October 2024), there are no concrete data yet as to the impact on Egypt's water supply. According to the Egyptian Foreign Ministry, however, even a 2 percent reduction of water supply could lead to the loss of 200,000 acres of irrigated land.[21] In any case, the full impact of the dam is expected to take several years to materialize. What is clear, however, is that the dam could plunge the country further into water poverty. As of 2022, the per capita water share has already declined to below 600 cubic meters annually, well short of the water poverty line. These serious challenges persist despite the resurgent hope of finding immense reservoirs of water underneath Egyptian soil.[22] Indeed, while studies suggest that the Nubian Sandstone Aquifer System—located under parts of Egypt, Sudan, Libya, and Chad—is the world's largest known aquifer, only a relatively small volume of water is economically extractable.

At the same time, climate change is likely to intensify existing conflicts over water resources. The increasing frequency and severity of extreme weather events, including droughts, storms, and floods, are placing immense stress on urban water systems. For millions of city dwellers around the world, water scarcity has become a daily reality, not a distant possibility happening in a future apocalypse.[23] Over the past decades, several major cities—including São Paulo, Mexico City, Cape Town, and Tehran—have announced that they were dangerously close to "Day Zero," the anticipated date when taps would run dry. Invariably, however, poor management of water supplies is always a contributing factor in water crises, shaping their social distribution.[24] Whereas Egyptian cities are so far not severely affected by global warming, any future planning has to consider these eventualities.

During the last few years, the Egyptian government has taken steps to reduce the country's vulnerability to external threats and improve the overall management of water resources. Toward this end, desalination and wastewater treatment projects have been significantly expanded, while the cultivation of water-intensive crops, such as rice, was reduced. At the same time, however, Egypt has moved further into water privatization, with the holding company introducing a new type of prepaid meters—forcing users to pay before consumption—which has led to more resistance from the public. As one study shows, Cairo's inhabitants are unwilling to pay for water services not because of the irrational belief that water is a "gift" from the divine, but because of their direct experience with the dysfunctional and arbitrary payment system.[25] Another study shows that the turn toward privatization has not resulted in a more efficient and rational service, but rather has produced widespread informality in both informal and elite areas.[26]

In September 2025, the government issued a law that allows for the complete privatization of the water sector.[27] For the first time since 1957, private companies with full independence from the state could supply Nile water to Cairo and other Egyptian cities—a red line that had not previously been crossed. Unlike the holding company, these firms could set pricing policies and investment strategies, issue shares to be traded on the stock market, and merge or liquidate with minimal oversight from government agencies. Without active monitoring by municipal bodies and Parliament, private companies would enjoy a degree of power and autonomy resembling that of the old concessionary corporation which is the subject of this book.

This latest decision is all the more jarring considering the negative experience of water privatization in major world cities, where it was found to have led to higher prices, environmental pollution, and uneven investment in water infrastructure. Therefore, the past two decades have seen a growing trend of cities retaking water services under public management—with at least 267 cases of water remunicipalization in thirty-seven countries.[28] In Paris, for instance, water distribution was brought back under public municipal management in 2010, after a twenty-five-year period of privatization. Supporters of this move argued that it would improve transparency, reinvest profits into infrastructure, ensure sustainable water management, lower costs for consumers, and treat water as a public good rather than a commodity.[29] In Argentina, Buenos Aires's water supply was brought back under public ownership after a brief period of privatization (1993–2006), having been found to lead to unequitable access and inefficient treatment of water contamination.[30] In London, where water provisioning is currently under private control, the water companies have come under intense public scrutiny after it was revealed that they disinvested in infrastructure while still raising the price on consumers by 40 percent and distributing large profits to shareholders.[31]

Irrespective of the lessons to be learned from distant and recent history, the political economy of water management will remain, in essence, a site of political struggle. It is the struggle between progressive forces calling for public control of common resources—not as a step toward a future cornucopia, but as a necessary measure for any real assessment of society's needs and capabilities—and conservative tendencies that see nothing in common except the maximizing economic agent, whose ultimate expression is found in the capitalist world economy.

NOTES

Introduction

1. Davison, "The City as a Natural System"; Cronon, *Nature's Metropolis*; Harvey, *Paris*, 88–112; Melosi, "The Place of the City in Environmental History"; Pike, *Subterranean Cities*.

2. Hugo, *Les Misérables*, 1130. I have taken some liberty to modify the translation. The French *cloaques* has been rendered as sewers (in Isabel Hapgood's translation), cloaca (by Norman Denny), and waste dumps (by Christine Donougher).

3. Wolmar, *Blood, Iron, and Gold*.

4. Phalen, *How the Telegraph Changed the World*.

5. Shadwell, *The London Water Supply*, 264–68; Kellet, "Municipal Socialism"; Millward, "Emergence of Gas and Water Monopolies"; Millward, "The Political Economy of Urban Utilities"; Broich, *London*, ch. 3; Stromquist, "Municipal Socialism."

6. Reliable data records are mostly unavailable for this period, but some information could be found in: Hodge, *Roman Aqueducts*; Magnusson, *Water Technology in the Middle Ages*; Mays, *Ancient Water Technologies*.

7. Crouch et al., "Defining Domestic Water Consumption Based on Personal Water Use Activities"; Dieter and Maupin, *Public Supply*.

8. Hugo, *Les Misérables*, 1137–38.

9. Graham-Leigh, *London's Water Wars*; Broich, *London*.

10. Laporte, *History of Shit*; Corbin, *The Foul and the Fragrant*; Forty, *Objects of Desire*, ch. 7; Hamlin, *Public Health in the Age of Chadwick*; Rogaski, *Hygienic Modernity*.

11. Trentmann and Taylor, "From Users to Consumers"; Taylor and Trentmann, "Liquid Politics"; Broich, *London*; Bocquet, Chatzis, and Sander, "From

Free Good to Commodity"; Björkman, *Pipe Politics*; Trentmann, *Empire of Things*, ch. 3; Anand, *Hydraulic City*.

12. Reid, *Paris Sewers and Sewermen*.

13. Tarr, "The Evolution of the Urban Infrastructure"; Graham and Marvin, *Splintering Urbanism*.

14. Goubert, *The Conquest of Water*; Melosi, *The Sanitary City*; Gandy, *The Fabric of Space*; Hamlin, "Edwin Chadwick and the Engineers."

15. For studies on the fossil fuel complex see Mitchell, *Carbon Democracy*; Malm, *Fossil Capital*; Barak, *Powering Empire*. On Barak is exceptional in treating water as a critical component of changing energy regimes.

16. Landes, *Bankers and Pashas*; Piquet, "The Suez Company's Concession in Egypt"; Hubert, *History of the Suez Canal Company*.

17. Crouchley, *The Investment of Foreign Capital in Egyptian Companies*, 35–36.

18. Arnaud, *Le Caire*, 31–80.

19. Howard-Jones, *The Scientific Background of the International Sanitary Conferences*. See also Fahmy, "An Olfactory Tale of Two Cities"; Fahmy, "Modernizing Cairo"; Fahmy, *In Quest of Justice*, ch. 3.

20. Owen, "The Cairo Building Industry"; Jakes, *Egypt's Occupation*.

21. Pyramid, "The Drainage of Cairo," 301.

22. DWQ, Diwan al-Ashghal al-'Umumiyya (Public Works Ministry), file no. 4003-023470, From Elwin Palmer to Greene Pasha, September 4 (1890). The Egyptian pound was almost equivalent to the British pound: L.E. 1 = £. 1. 0s. 6d / L.E. 0.975 = £. 1.

23. Public Works Ministry, *Report Upon the Administration of the Public Works Department for 1898*, 186.

24. Wasfi, *Muhadara 'an Majari al-Qahira*, 27.

25. This is a simplified version of G. F. C. Rogers's definition in *The Nature of Engineering*, 51. See also Vincenti, *What Engineers Know*; Ferguson, *Engineering and the Mind's Eye*; Koen, *Discussion of the Method*; Madhavan, *Applied Minds*.

26. My review of the engineering literature excludes Whig accounts such as Samuel Smiles's *Lives of the Engineers* (1862), which lionizes engineers as the "makers of modern civilization" and a driving force behind British imperial dominance. See Briggs, "Samuel Smiles and the Gospel of Work." For a different take on the historiography, see Picon, "Engineers and Engineering History."

27. Layton, *The Revolt of the Engineers*; Moore, *Images of Development*; Kranakis, "Social Determinants of Engineering Practice"; Buchanan, *The Engineers*; Picon, *L'Invention de l'ingénieur moderne*; Oldenziel, *Making Technology Masculine*; Wisnioski, *Engineers for Change*; Bix, *Girls Coming to Tech*.

28. Noble, *America by Design*; Alder, *Engineering the Revolution*; Camprubí, *Engineers*; Derr, *The Lived Nile*, chs. 1–2; Björkman and Harris, "Engineering Cities." See also Barak, *On Time*; Mieton, *Electrical Palestine*; Stamatopoulou-Robbins, *Waste Siege*; Günel, *Spaceship in the Desert*; Khalili, *Sinews of War and Trade*.

29. Callon, "Society in the Making"; Law and Callon, "Engineering and Sociology"; Vincenti, *What Engineers Know*; Porter, "Objectivity and Authority"; Porter, *Trust in Numbers*, ch. 3; Latour, *Aramis*; Mitchell, *Rule of Experts*, ch. 3; Mukerji, *Impossible Engineering*; Picon, "The Engineer as Judge"; Mitcham, *Thinking Through Technology*; Voskuhl, "Engineering Philosophy"; Duarte and Giraud, *Economics and Engineering*.

30. Bijker, Hughes, and Pinch, *Social Construction of Technological Systems*; Shapin and Schaffer, *Leviathan and the Air-pump*; Schaffer, "Babbage's Intelligence"; Light, "When Computers were Women."

31. Lardner, *Railway Economy*.

32. Etner, *Histoire du calcul économique en France*; Porter, "Objectivity and Authority"; Porter, *Trust in Numbers*, ch. 3; Ekelund and Hébert, *Secret Origins of Modern Microeconomics*; Ekelund and Hébert, "Retrospectives."

33. Veblen, *The Engineers and the Price System*. See also: Stabile, "Veblen and the Political Economy of the Engineer"; Knoedler and Mayhew, "Thorstein Veblen and the Engineers."

34. Perelman, *Railroading Economics*, Introduction and ch. 1.

35. Marx, *Capital*, Vol. 2, 288.

36. Schwartz, "John Stuart Mill and Laissez Faire"; Tynan, "Mill and Senior"; Mosca, "On the Origins of the Concept of Natural Monopoly."

Chapter 1

1. Hunter, *Egypt Under the Khedives*, 165–76; Goldschmidt, *Biographical Dictionary*, 157–58.

2. Brown, "The Precarious Life and Slow Death of the Mixed Courts of Egypt."

3. Hunter, *Egypt Under the Khedives*, 165; Goldschmidt, *Biographical Dictionary*, 158.

4. Hunter, *Egypt Under the Khedives*, 170.

5. Nubar, *Mudhakirat Nubar Pasha*, 354.

6. Crouchley, *The Investment of Foreign Capital*, 35–36. Pasquier, *La Société des Eaux du Caire*, 117–18.

7. Saul, *La France et l'Égypte*, 40–44; Pasquier, *La Société des Eaux du Caire*, 130–43.

8. Al-Rafi'i, *'Asr Imsa'il*; Owen, *The Middle East in the World Economy*, ch. 5.

9. Marx, *Capital*, Vol. 1, 873–95; Harvey, *Rebel Cities*, 67–88.

10. Howard-Jones, *The Scientific Background of the International Sanitary Conferences*.

11. Fahmy, "An Olfactory Tale of Two Cities"; Fahmy, "Modernizing Cairo"; Fahmy, *In Quest of Justice*, ch. 3.

12. Al-Maqrizi, *Al-Maw'aiz wa al-I'tibar*, Vol. 2, 758–59.

13. Al-Maqrizi, *Al-Maw'aiz wa al-I'tibar*, Vol. 1, 175–90; Kaufmann, "Sur le Prétendu Nil Vert," 105–13.

14. Beinin and Lockman, *Workers on the Nile*, ch. 2; 'Amir, *Tawa'if al-Hiraf fi Misr*; Chalcraft, *The Striking Cabbies of Cairo*, ch. 6; Al-Tukhi, *Tawa'if al-Hiraf.*

15. Amin, *Qamus al-'Adat wa al-Taqalid wa al-Ta'abir al-Misriyya*, 235; Raymond, *Cairo*, 245–47; Al-Tukhi, *Tawa'if al-Hiraf*, 87–90.

16. Raymond, *Cairo*, 241–50; Hanna, *An Urban History of Būlāq*.

17. 'Afifi, *Al-Awqaf wa al-Haya al-Iqtisadiyya fi Misr.*

18. Pauty, *Les Hammams du Caire*, 4.

19. For instance, see Commission des Sciences et des Arts, *Description De l'Égypte*, Vol. 10, 218–21.

20. Commission des Sciences et des Arts, *Description De l'Égypte*, Vol. 10, 220; [Adolphe Smith], "The Cleansing of Cairo (From Our Special Sanitary Commissioner)," *The Lancet*, June 12, 1909.

21. Fadli and Sibley, "The Historic *Hammāms* of Cairo."

22. [Adolphe Smith], "The Cleansing of Cairo (From Our Special Sanitary Commissioner)," *The Lancet*, June 12, 1909.

23. For more on this construction material (*qusrumil*), see Wasfi, *Al-Qawa'id al-Asasiyya*, Vol. 1, 25.

24. Mostafa, "The Cairene Sabil"; Dobrowolska and Dobrowolski, *The Sultan's Fountain.*

25. Al-Husayni, *Al-Asbila al-'Uthmaniyya*, 42.

26. Raymond, *Cairo*, 246; Mubarak, *Al-Khitat al-Tawfiqiyya*, Vol. 1, 83.

27. Al-Husayni, *Al-Asbila al-'Uthmaniyya*, 71–76.

28. Mubarak, *Al-Khitat al-Tawfiqiyya*, Vol. 1, 87; "Rapport de la commission internationale de l'assainissement du Caire," *Journal Officiel*, Supplement to no. 51 (16 April, 1892).

29. DWQ, Diwan al-Ashghal al-'Umumiyya (Public Works Ministry), file no. 4003-037344, "Rapport de la commission d'assainissement du Caire" (1885).

30. Al-Maqrizi, *al-Mawa'iz wa al-I'tibar*, Vol. 1, 258–59.

31. Wasfi, *Al-Qawa'id al-Asasiyya*, Vol. 2, 4; Said, *Nineteenth Century Houses*, chs. 2–3.

32. Mahmud Mustafa, "Miyah al-Abar bi al-Mahrusa," *al-Sihha* 1, no. 2 (September, 1887), 61; Wasfi, *Al-Qawa'id al-Asasiyya*, Vol. 2, 23; Clerget, *Le Caire*, Vol. 1, 54–55.

33. Wasfi, *Al-Qawa'id al-Asasiyya*, Vol. 2, 7; Said, *Nineteenth Century Houses*, 71–74.

34. On water privatization since the 1970s, see Graham and Marvin, *Splintering Urbanism*; Swyngedouw, Kaïka, and Castro, "Urban Water"; Castree, "Neoliberalising Nature"; Bakker, *Privatizing Water.*

35. Graham-Leigh, *London's Water Wars*, chs. 3 and 5.

36. Mill, *Essays on Economics and Society*, Part II, 433–37; Schwartz, "John Stuart Mill and Laissez Faire"; Winch, "The Problematic Status of the Consumer in Orthodox Economic Thought," 37.

37. Stigler, "The Economists and the Problem of Monopoly"; Mosca, "On the Origins of the Concept of Natural Monopoly."

38. Schwartz, "John Stuart Mill and Laissez Faire"; Kellet, "Municipal Socialism"; Millward, "Emergence of Gas and Water Monopolies"; Millward, "The Political Economy of Urban Utilities"; Broich, *London*, ch. 3.

39. Pasquier, *La Société des Eaux du Caire*, 133–36.

40. *Firman autorisant l'établissement d'une distribution d'eau du nil au Caire et ses faubourgs* [Firman Authorizing the Establishment of a Nile Water Distribution System in Cairo and Its Suburbs], in Ministère des Finances, *Recueil Général de Contrats*, Article 2.

41. *Firman autorisant l'établissement d'une distribution d'eau du nil au Caire et ses faubourgs* [Firman Authorizing the Establishment of a Nile Water Distribution System in Cairo and Its Suburbs], in Ministère des Finances, *Recueil Général de Contrats*, Article 5.

42. Forty paras = one piaster = ten millimes.

43. *Firman autorisant l'établissement d'une distribution d'eau du nil au Caire et ses faubourgs* [Firman Authorizing the Establishment of a Nile Water Distribution System in Cairo and Its Suburbs], in Ministère des Finances, *Recueil Général de Contrats*, Article 5.

44. Ibid.

45. DWQ, Nizarat al-Dakhiliyya (Ministry of Interior), file no. 2001-009559, Various correspondences concerning the Cairo Water Company's exemption from customs duties (1876); and file no. 2001-0096112, Contract for watering the streets between the Cairo Water Company and the Egyptian Government (1878).

46. Al-Jabarti, *'Aja'ib al-Athar*, Vol. 4, 255; Bret, "Gunpowder Manufacture in Cairo."

47. Kuhnke, *Lives at Risk*, 65–68. See also Nubar, *Mudhakirat Nubar Pasha*, 363–67.

48. Abu-Lughod, *Cairo*, ch. 7.

49. Abu-Lughod, *Cairo*, 105–11; Hunter, *Egypt Under the Khedives*, 123–38; Reimer, "Contradiction and Consciousness"; Arnaud, *Le Caire*, 81–83.

50. Mubarak, *Al-Khitat al-Tawfiqiyya*, Vol. 1, 83; Berque, *Egypt*, 92; Reid, *Whose Pharaohs?*, 230–34.

51. See the lists of names of places, roads, squares, and establishments in the Cairo survey published in: Commission des Sciences et des Arts, *Description De l'Égypte*, Vol. 10.

52. Mubarak, *Al-Khitat al-Tawfiqiyya*, Vol. 1, 80.

53. Arnaud, *Le Caire*, 31–80.

54. Ibid., 50–51.

55. Mubarak, *Al-Khitat al-Tawfiqiyya*, Vol. 9, 56.

56. DWQ, Muhafazat Misr (Cairo Province), file no. 2002-004293 and 2002-004294, Various correspondences between Grand Bey and the director of Cairo

Province concerning land expropriation at Shaykh Rihan Hill (1871); DWQ, Muhafazat Misr (Cairo Province), file no. 2002-008490 to 2002-008550, Various correspondences concerning al-'Abit Village (1871).

57. Arnaud, *Le* Caire, 54–75.

58. Schölch, *Egypt for the Egyptians*; Cole, *Colonialism and Revolution.*

59. British Foreign Office, *Correspondences respecting the cholera epidemic in Egypt: 1883*, 92, "The Outbreak of Cholera at Damietta in 1883: Its Origin and Spread. Report to the Maritime Sanitary and Quarantine Board of Egypt by Ahmed Chaffey [Shafi'i] Bey and Salvatore Ferrari, July 24, 1883."

60. British Foreign Office, *Further correspondences respecting the cholera epidemic in Egypt: 1883*, 51–52, "Statement Showing the Mortality from Cholera in Cairo and the Suburb of Ghizeh [Giza], from July 15 to August 25, inclusive"; Ministry of Interior, Sanitary Department, *Report on the Epidemic of Cholera in Egypt During the Years 1895 & 1896*, 60–61; Bitter, "Le choléra et la peste en Égypte," 215.

61. Howard-Jones, *The Scientific Background of the International Sanitary Conferences*, 23–34; Kuhnke, *Lives at Risk*, ch. 5.

62. The Board was led by Hasan Mahmud. See "Doctor Hasan Pasha Mahmud," *Al-Majalla al-Tibbiyya al-Misriyya* 1, no. 3 (1917): 2–7.

63. It was led by Salim Salim. See "Doctor Salim Pasha Salim," *Al-Muqtataf* 18, no. 4 (1894): 217–19.

64. British Foreign Office, *Further correspondences respecting the cholera epidemic in Egypt: 1883*, 50, "Messrs. Gibson and Ornstein to Sir E. Mallet, August 25, 1883."

65. Davis, *Planet of Slums*, 21–22.

66. Ackerknecht, "Anticontagionism between 1821 and 1867."

67. British Foreign Office, *Correspondences respecting the cholera epidemic in Egypt: 1883*, 87–95, "The Outbreak of Cholera at Damietta in 1883: Its Origin and Spread. Report to the Maritime Sanitary and Quarantine Board of Egypt by Ahmed Chaffey [Shafi'i] Bey and Salvatore Ferrari, July 24, 1883."

68. British Foreign Office, *Further correspondences respecting the cholera epidemic in Egypt: 1883*, 28, "Consul Cookson to Earl Granville, August 16, 1883." On disease and rumors, see Briggs, "Cholera and Society in the Nineteenth Century" and White, *Speaking with Vampires*, ch. 2.

69. Watts, "From Rapid Change to Stasis"; Low, *Imperial Mecca*, ch. 3.

70. Coleman, "Koch's comma bacillus"; Ogawa, "Uneasy Bedfellows."

71. TNA, FO 407/32, From Evelyn Baring to Earl Granville, November 5 (1883).

72. See Grant's account of his dismissal in Grant, "Cholera"; Watts, "From Rapid Change to Stasis," 360–63.

73. Egyptian Government, *Qararat wa Manshurat*, 43–44, "Interior Ministry circular regarding the Council of Ministers' decision on the plan to construct waterproof cesspools and the need to replace mosque ablution basins with faucets," May 7 (1884); DWQ, Diwan al-Ashghal al-'Umumiyya (Public Works

Ministry), file no. 4003-037345, Various correspondences between the Ministry of Interior and the *Awqaf*, May 28 and December 17 (1884); Third and fourth sessions of the Cairo Sanitary Commission (*commission d'assainissement du Caire*), held on February 7 and 12 (1885).

74. Lane-Poole, *The Story of Cairo*, viii.

75. TNA, FO 141/581, Response of William Willcocks to a questionnaire by the Milner mission of enquiry, November 25 (1919).

76. Quoted in Hollings, *The Life of Sir Collin C. Scott-Moncrieff*, 176.

77. DWQ, Diwan al-Ashghal al-'Umumiyya (Public Works Ministry), file no. 4003-037345, Sessions of the Cairo Sanitary Commission (1885).

78. Fahmy, *In Quest of Justice*, ch. 3.

79. DWQ, Diwan al-Ashghal al-'Umumiyya (Public Works Ministry), file no. 4003-037345, First session of the Cairo Sanitary Commission, held on January 31 (1885).

80. DWQ, Diwan al-Ashghal al-'Umumiyya (Public Works Ministry), file no. 4003-037345, "Taqrir Yakhuss Khalij al-Qahira 'ala Wajh al-Ikhtisar [Summary Report on the Cairo Khalij]," by Latif Salim (1885); Third session of the Cairo Sanitary Commission, held on February 7 (1885).

81. DWQ, Diwan al-Ashghal al-'Umumiyya (Public Works Ministry), file no. 4003-037345, Note by Scott-Moncrieff (read by Julien Barois) in the tenth and final session of the Cairo Sanitary Commission, held on June 4 (1885).

82. DWQ, Diwan al-Ashghal al-'Umumiyya (Public Works Ministry), file no. 4003-037344, "Rapport de la commission d'assainissement du Caire" (1885).

83. Abbate, "Questions hygiéniques," 63.

84. DWQ, Diwan al-Ashghal al-'Umumiyya (Public Works Ministry), file no. 4003-037344, "Rapport de la commission d'assainissement du Caire" (1885).

85. DWQ, Diwan al-Ashghal al-'Umumiyya (Public Works Ministry), file no. 4003-037345, Seventh session of the Cairo Sanitary Commission, held on March 28 (1885).

86. DWQ, Diwan al-Ashghal al-'Umumiyya (Public Works Ministry), file no. 4003-037344, "Rapport de la commission d'assainissement du Caire" (1885).

87. DWQ, Diwan al-Ashghal al-'Umumiyya (Public Works Ministry), file no. 4003-037345, Memo by Greene Pasha, director of the Department of Public Health and Sanitary Services, November 12 (1887); Two field reports on conditions of present sewers (1886–87).

88. DWQ, Nizarat al-Dakhiliyya (Ministry of Interior), file no. 2001-018157, "Ministerial Decree Concerning the Construction of Workmen's Dwellings," October (1883).

89. DWQ, Diwan al-Ashghal al-'Umumiyya (Public Works Ministry), file no. 4003-037344, "Rapport de la commission d'assainissement du Caire" (1885).

90. DWQ, Diwan al-Ashghal al-'Umumiyya (Public Works Ministry), file no. 4003-037344, "Rapport de la commission d'assainissement du Caire" (1885).

91. DWQ, Majlis al-Wuzara' (Council of Ministers), file no. 0075-008831, Contract between the Egyptian Government and the Cairo Sewage Transport Company, August (1886).

92. Clerget, *Le Caire*, Vol. 2, 4.

93. *Convention prorogeant jusqu'au 1er Janvier 1938 la faculté réservée au Gouvernement pour le rachat de la concession* [Convention Extending until January 1, 1938, the Power Reserved to the Government to Redeem the Concession], January 27 (1887), in Ministère des Finances, *Recueil Général de Contrats*, Article 6.

94. Goubert, *The Conquest of Water*, 188.

95. Shadwell, *The London Water Supply*, 66–69.

96. Pyramid, "The Drainage of Cairo," 301.

97. DWQ, Majlis al-Wuzara' (Council of Ministers), file no. 0075-059295, "Report to his Excellency the Minister of Public Works on the Charges for Water in Cairo," by S. R. Raffety, February 4 (1938).

98. *Procès-verbal de l'assemblée générale extraordinaire de la Société Anonyme des Eaux du Caire approvant la convention du 27 Janvier 1887* [Minutes of the Extraordinary General Meeting of the Cairo Water Company Approving the Convention of January 27, 1887], February 22 (1887), in Ministère des Finances, *Recueil Général de Contrats.*

99. Crouchley, *The Investment of Foreign Capital*, 36.

100. *Tarif Général pour la ville du Caire* [General Tariff for the City of Cairo], January 1 (1887), in Ministère des Finances, *Recueil Général de Contrats.*

101. *Rachat de la participation du Gouvernement dans les bénéfices de la Société Anonyme des Eaux du Caire* [Repurchase of the Government's Participation in the Profits of the Cairo Water Company], July 10 (1889), in Ministère des Finances, *Recueil Général de Contrats.*

102. *Procès-verbal de l'assemblée générale extraordinaire de la Société Anonyme des Eaux du Caire approvant la convention du 10 Juillet 1889* [Minutes of the Extraordinary General Meeting of the Cairo Water Company Approving the Convention of January 27, 1887], July 10 (1889), in Ministère des Finances, *Recueil Général de Contrats.*

Chapter 2

1. Milner, *England in Egypt*, ch. 8.

2. Tignor, *Modernization and British Colonial Rule in Egypt*, chs. 3–4.

3. Ministry of Finance, Statistical Department, *Statistical Yearbook of Egypt for 1909*, 51.

4. Brown, *History of the Barrage*, 40.

5. DWQ, Diwan al-Ashghal al-'Umumiyya (Public Works Ministry), file no. 4003-037344, Papers of the review commission, February 10 (1890). Scott-Moncrieff and Price were the ones behind the idea of installing faucets inside mosques a few years earlier (Chapter 1).

6. DWQ, Diwan al-Ashghal al-'Umumiyya (Public Works Ministry), file no. 4003-037344, Notes on Cairo drainage by A. C. Reid, February 21 (1890).

7. DWQ, Diwan al-Ashghal al-'Umumiyya (Public Works Ministry), file no. 4003-023470, Draft report signed by Colin Scott-Moncrieff, April 17 (1890).

8. DWQ, Diwan al-Ashghal al-'Umumiyya (Public Works Ministry), file no. 4003-021820, "Mr. Baldwin Latham's Reply to Mr. A. C. Reid's Notes on the Drainage of Cairo," March 19 (1890).

9. Watson, *Sewerage Systems.*

10. Foster, "Marx's Theory of Metabolic Rift."

11. Hanley, "Urban Sanitation in Preindustrial Japan."

12. Ibid., 9–11.

13. Foster, "Marx's Theory of Metabolic Rift," 375.

14. Ibid., 375–76.

15. Ibid., 377; Smil, *Enriching the Earth,* 40–43; Cushman, *Guano.*

16. Mårald, "Everything Circulates"; Cooper, "Peter Lund Simmonds."

17. Simmons, "Waste Not, Want Not."

18. Liebig, *Letters on Modern Agriculture.*

19. Quoted in Foster, "Marx's Theory of Metabolic Rift," 377.

20. Quoted in Saito, *Karl Marx's Ecosocialism,* 198. See also Liebig, *Two Letters.*

21. Hugo, *Les Misérables,* 1126–29.

22. Foster, "Marx's Theory of Metabolic Rift"; Saito, *Karl Marx's Ecosocialism.*

23. Quoted in Saito, *Karl Marx's Ecosocialism,* 292.

24. Marx, *Capital,* Vol. 3, 949.

25. Ibid., Vol. 1, 637–38.

26. Corbin, *The Foul and the Fragrant,* 119–20; Reid, *Paris Sewers and Sewermen,* ch. 6; Gandy, "The Paris Sewers and the Rationalization of Urban Space"; Tarr, "The Separate vs. Combined Sewer Problem."

27. DWQ, Diwan al-Ashghal al-'Umumiyya (Public Works Ministry), file no. 4003-037344, Note by Dr. Abbate, April 20 (1890).

28. Institution of Civil Engineers Archives, United Kingdom, Baldwin Latham papers.

29. W., "Obituary," 100.

30. Hamlin, *Public Health in the Age of Chadwick*; Hamlin, "Edwin Chadwick and the Engineers."

31. Public Works Ministry, *Résumé du rapport de M. Baldwin Latham sur le drainage du Caire.*

32. Estimated at £. 47,500.

33. Public Works Ministry, *Résumé du rapport de M. Baldwin Latham.* Bulaq and Old Cairo were not considered part of Cairo proper. The total number of buildings including the two suburbs was 52,000.

34. DWQ, Diwan al-Ashghal al-'Umumiyya (Public Works Ministry), file no. 4003-037344, Note by Dr. Abbate, April 20 (1890).

35. Howard-Jones, *The Scientific Background of the International Sanitary Conferences*, 35–41.

36. DWQ, Diwan al-Ashghal al-'Umumiyya (Public Works Ministry), file no. 4003-037344, Note by Dr. Abbate, April 20 (1890).

37. Scott, "The Liernur Sewerage System at Dordrecht (Holland)," 669.

38. Watson, *Sewerage Systems*, 5–6.

39. Scott, "The Liernur Sewerage System at Dordrecht (Holland)"; Mårald, "Everything Circulates."

40. DWQ, Diwan al-Ashghal al-'Umumiyya (Public Works Ministry), file no. 4003-021820, "Mr. Baldwin Latham's Reply to Mr. A. C. Reid's Notes on the Drainage of Cairo," March 19 (1890).

41. DWQ, Diwan al-Ashghal al-'Umumiyya (Public Works Ministry), file no. 4003-023470, Draft report signed by Colin Scott-Moncrieff, April 17 (1890).

42. Ellet, *An Essay on the Laws of Trade*; Ellet, *The Laws of Trade*; Lardner, *Railway Economy*; Dupuit, *De L'Utilité et de sa Mesure*.

43. Ellet, *An Essay on the Laws of Trade*, vii.

44. Ellet, *The Laws of Trade*, 15.

45. Ibid.

46. Ekelund and Hébert, *Secret Origins of Modern Microeconomics*, ch. 7.

47. Quinn and Turner, *Boom and Bust*, ch. 4.

48. Lardner, *Railway Economy*, 6–7.

49. Odlyzko, "Dionysius Lardner"; Odlyzko, "Collective Hallucinations."

50. Lardner, *Railway Economy*, ch. 13.

51. Ibid., 287–89.

52. Porter, "Objectivity and Authority"; Porter, *Trust in Numbers*, 57–59; Ekelund and Hébert, *Secret Origins of Modern Microeconomics*, ch. 2.

53. Smith, "The Longest Run."

54. Ekelund and Hébert, *Secret Origins of Modern Microeconomics*, 75.

55. Dupuit, "On the Measurement of the Utility of Public Works"; Dupuit, "On Tolls and Transportation Charges."

56. Dupuit, *De L'Utilité et de sa Mesure*, 174–75.

57. Dupuit, "On the Measurement of the Utility of Public Works," 261.

58. Ekelund and Hébert, *Secret Origins of Modern Microeconomics*, ch. 13.

59. Dobb, *Theories of Value and Distribution since Adam Smith*, ch. 7; Black, "W. S. Jevons and the Foundation of Modern Economics"; Heilbroner and Milberg, *Crisis of Vision*, chs. 1–2; Milonakis and Fine, *From Political Economy to Economics*; Kurz, *Economic Thought*, ch. 4.

60. Jevons, *The Theory of Political Economy*, Preface; Jevons, *Letters and Journals*, 384.

61. Public Works Ministry, *Résumé du rapport de M. Baldwin Latham.*

62. Ibid.

63. Latham, *Sanitary Engineering*, chs. 9–10.

64. 1 Imperial Gallon = 4.5 Liters = 0.0045 Cubic Meters.

65. Public Works Ministry, *Résumé du rapport de M. Baldwin Latham.*

66. Ibid.

67. DWQ, Diwan al-Ashghal al-ʻUmumiyya (Public Works Ministry), file no. 4003-037344, Notes on Cairo drainage by A. C. Reid, February 21 (1890).

68. Ibid.

69. Ibid.

70. Ibid.

71. See Chapter 4 for later studies on living expenses. The British colonial government took up this question after the 1919 revolution. See TNA, FO 848/4 (Milner Mission), Memorandum by Dr. W. H. Wilson on Cost of Living to the Poorer Classes in Egypt (1919) and Ministry of Finance, Statistical Department, *Report on Living Expenses*. See also Labib, *Recenser l'Égypte*.

72. DWQ, Nizarat al-Dakhiliyya (Ministry of Interior), file no. 2001-018157, Concerning the deplorable conditions of slums (workmen's dwellings), (1889).

73. DWQ, Diwan al-Ashghal al-ʻUmumiyya (Public Works Ministry), file no. 4003-037344, Notes on Cairo drainage by John Price, April 1 (1890).

74. DWQ, Diwan al-Ashghal al-ʻUmumiyya (Public Works Ministry), file no. 4003-023470, Draft report signed by Colin Scott-Moncrieff, April 17 (1890).

75. Ibid.

76. DWQ, Diwan al-Ashghal al-ʻUmumiyya (Public Works Ministry), file no. 4003-021820, "Mr. Baldwin Latham's Reply to Mr. A. C. Reid's Notes on the Drainage of Cairo," March 19 (1890).

77. Ibid.

78. Ibid.

79. Ibid.

80. Ibid.

81. Ibid.

82. Ibid.

83. Ibid.

84. Layton, *The Revolt of the Engineers*; Kranakis, "Fixing the Blame."

85. DWQ, Diwan al-Ashghal al-ʻUmumiyya (Public Works Ministry), file no. 4003-023470, From Elwin Palmer to Greene Pasha, September 4 (1890).

86. Tunçer, *Sovereign Debt and International Financial Control*, 44–49.

87. France, Ministère des Affaires Étrangères, *Documents Diplomatiques*, 446.

88. Scott, *The Law Affecting Foreigners in Egypt*.

89. The French government approved the tax two years after the London Convention (1885). See Todd, *Political Economy*, 344–45.

90. France, Ministère des Affaires Étrangères, *Documents Diplomatiques*, 446.

Chapter 3

1. Tignor, *Modernization and British Colonial Rule in Egypt,* 113.

2. Owen, *Cotton and the Egyptian Economy,* chs. 8–9; Berque, *Egypt,* ch. 4; Owen, *The Middle East in the World Economy,* ch. 9.

3. Derr, *The Lived Nile,* chs. 1–2.

4. Ministry of Finance, Survey of Egypt, *Geology of Egypt.*

5. Ministry of Finance, Survey Department, *The Cadastral Survey of Egypt,* 1–12.

6. Mitchell, *Rule of Experts,* 90.

7. Pyramid, "The Drainage of Cairo," 300.

8. DWQ, Diwan al-Ashghal al-'Umumiyya (Public Works Ministry), file no. 4003-037312 and 4003-037313, Newspaper clips, November–December (1891).

9. "Rapport de la commission internationale de l'assainissement du Caire," *Journal Officiel,* supplement to no. 51 (April 16, 1892), 1.

10. Sutcliffe, *Towards the Planned City,* 20–22; Bernet, "The 'Hobrecht Plan' (1862) and Berlin's Urban Structure."

11. Hobrecht, *Die Canalisation von Berlin.*

12. Hastaoglou-Martinides, "The building of Istanbul docks," 92.

13. DWQ, Diwan al-Ashghal al-'Umumiyya (Public Works Ministry), file no. 4003-037313, Requests for information (1891–92).

14. DWQ, Diwan al-Ashghal al-'Umumiyya (Public Works Ministry), file no. 4003-037313, From J. L. Michau & Fils to Muhammad Zaki Pasha (Minister of Public Works), March 20 (1890).

15. DWQ, Diwan al-Ashghal al-'Umumiyya (Public Works Ministry), file no. 4003-021820, Note by Colin Scott-Moncrieff, April 30 (1890).

16. DWQ, Diwan al-Ashghal al-'Umumiyya (Public Works Ministry), file no. 4003-037313, Correspondence between Isaac Shone and the Egyptian Public Works Ministry, December 4 (1891) to January 8 (1892).

17. Shone, *The Main Drainage of the Houses of Parliament.*

18. DWQ, Diwan al-Ashghal al-'Umumiyya (Public Works Ministry), file no. 4003-037313, Papers of the International Sanitary Commission (1892).

19. "Rapport de la commission internationale de l'assainissement du Caire," *Journal Officiel,* supplement to no. 51 (April 16, 1892), 2.

20. Ibid.

21. DWQ, Diwan al-Ashghal al-'Umumiyya (Public Works Ministry), file no. 4003-037312, Note by Julien Barois on Cairo drainage, July 10 (1891).

22. "Rapport de la commission internationale de l'assainissement du Caire," *Journal Officiel,* supplement to no. 51 (April 16, 1892), 10.

23. Ibid.

24. Ibid., 11.

25. Ibid., 14.

26. Ibid., 10.

27. Barois, *Rapport sur les recherches du pétrole de la Mer Rouge*; Barois, "Notice sur le climat du Caire"; Barois, *Les irrigations en Égypte*.

28. DWQ, Diwan al-Ashghal al-'Umumiyya (Public Works Ministry), file no. 4003-030151, Memo by the Public Works Ministry, May 24 (1892).

29. DWQ, Majlis al-Wuzara' (Council of Ministers), file no. 0075-037423, "Rapport général sur les travaux du bureau d'études de l'assainissement du Caire," by Julien Barois, January 15 (1895); Poffandi, *Indicateur égyptien*, 5, 18.

30. It is likely that the Grand Bey map was later updated using cut and paste technique, but these updates were still insufficient. See the copy of Grand Bey's map at the American University in Cairo's Rare Books and Special Collections Library.

31. Todd, *Political Economy*, 343–44.

32. Arnaud, *Le Caire*, 213–15.

33. DWQ, Majlis al-Wuzara' (Council of Ministers), file no. 0075-037423, "Rapport général sur les travaux du bureau d'études de l'assainissement du Caire," by Julien Barois, January 15 (1895).

34. Ibid.

35. Clerget, *Le Caire*, Vol. 1, 46–48.

36. DWQ, Diwan al-Ashghal al-'Umumiyya (Public Works Ministry), file no. 4003-030152, Correspondence relating to the detailed study, March 29 (1892), and Correspondence relating to the detailed study, September–December (1893).

37. DWQ, Diwan al-Ashghal al-'Umumiyya (Public Works Ministry), file no. 4003-030152, Correspondence relating to the detailed study, May 16 and June 12 (1892).

38. DWQ, Diwan al-Ashghal al-'Umumiyya (Public Works Ministry), file no. 4003-030152, Correspondence relating to the detailed study, May (1892).

39. DWQ, Diwan al-Ashghal al-'Umumiyya (Public Works Ministry), file no. 4003-030152, Correspondence relating to the detailed study, June 29 (1892).

40. DWQ, Diwan al-Ashghal al-'Umumiyya (Public Works Ministry), file no. 4003-030152, Correspondence relating to the detailed study, June 20 (1892).

41. Sandwith, *Egypt as a Winter Resort*, 104–5; Greene Pasha, "Mokattam: A Sanitorium for Cairo," *The Lancet*, June 29, 1895.

42. DWQ, Diwan al-Ashghal al-'Umumiyya (Public Works Ministry), file no. 4003-037314, Petition to the Khedive, April 20 (1894).

43. Ibid.

44. Ibid.

45. M. D., "Masarif al-'Asima," *al-Mu'ayyad*, March 25, 1894.

46. Ibid.

47. DWQ, Diwan al-Ashghal al-'Umumiyya (Public Works Ministry), file no. 4003-037314, Letter from Wilfrid Blunt to Lord Cromer, March 16 (1894).

48. DWQ, Diwan al-Ashghal al-'Umumiyya (Public Works Ministry), file no. 4003-037314, From William Garstin to Julien Barois, March 28 and April 24 (1894).

49. DWQ, Majlis al-Wuzara' (Council of Ministers), file no. 0075-037423, "Rapport général sur les travaux du bureau d'études de l'assainissement du Caire," by Julien Barois, January 15 (1895).

50. Pyramid, "The Drainage of Cairo," 300.

51. Ibid.

52. DWQ, Diwan al-Ashghal al-'Umumiyya (Public Works Ministry), file no. 4003-030152, From John Rogers to Colin Scott-Moncrieff, May 29 (1892).

53. DWQ, Diwan al-Ashghal al-'Umumiyya (Public Works Ministry), file no. 4003-037306, Note on the draft law by Julien Barois, June 18 (1898).

54. "Rapport de la commission internationale de l'assainissement du Caire," *Journal Officiel*, supplement to no. 51 (April 16, 1892), 6.

55. Alleaume, "Hygiène publique et travaux publics."

56. Quoted in Alleaume, "Hygiène publique et travaux publics," 173.

57. TNA, FO 407/127, Lord Cromer to the Earl of Kimberley, June 27 (1894).

58. Ibid.

59. Ministry of Finance, Survey Department, *The Cadastral Survey of Egypt*, 1–12.

60. 1 feddan = 1.038 acres = 4200 square meters.

61. Mitchell, *Rule of Experts*, 86.

62. Jakes, *Egypt's Occupation*, 96–98.

63. Arnaud, *Le Caire*, 214–15.

64. Mitchell, *Rule of Experts*, 87–88.

65. Judd, "Report on a Series of Specimens of the Deposits of the Nile Delta," 213.

66. Ministère de L'Intérieur, Département de l'Hygiène Publique, *Les Eaux d'Égypte* by A. Azadian, 19.

67. Clerget, *Le Caire*, Vol. 1, 55.

68. Horner, "An Account of Some Recent Researches Near Cairo" (1855 and 1858).

69. Gold, "Ancient Egypt and the Geological Antiquity of Man."

70. H. G. Lyons, Preface to Ministry of Finance, Survey of Egypt, *Geology of Egypt*, xxiii.

71. "Rapport de la commission internationale de l'assainissement du Caire," *Journal Officiel*, supplement to no. 51 (April 16, 1892), 3.

72. Smith, "The Subsoil of Cairo."

73. Judd, "Report on a Series of Specimens of the Deposits of the Nile Delta, Obtained by the Recent Boring Operations" and Judd, "Second Report on a Series of Specimens of the Deposits of the Nile Delta."

74. Ministry of Finance, Survey Department, *The Chemistry of River Nile* by Alfred Lucas, 36.

75. Lyons, Preface to Ministry of Finance, Survey of Egypt, *Geology of Egypt*, xxix.

76. Quoted in Willcocks, *Egyptian Irrigation*, 2.

77. Ibid., 4.
78. Sa'id, *Tarikh al-Da'wa ila al-'Amiyya*, 100–109.
79. *Al-Azhar*, 1, no. 1 (January 1893): 1–7.
80. *Al-Ustath*, January 10, 1893, 503–4.
81. *Al-Muhandis*, 1, no. 7 (August, 1893), 17.
82. *Al-Muhandis*, 1, no. 2 (March, 1893), Unnumbered page at the beginning.
83. *Al-Azhar*, 1, no. 2 (February 1893): 15–33. For more on British education policy, see Tignor, *Modernization and British Colonial Rule in Egypt*, 319–48, and Derr, *The Lived Nile*, 34–35.
84. Willcocks, *Sixty Years in the East*, 116–17, 142–43.
85. Howard-Jones, *The Scientific Background of the International Sanitary Conferences*, 66–70; Evans, *Death in Hamburg*.
86. Ministry of Interior, Sanitary Department, *Report on the Epidemic of Cholera in Egypt During the Years 1895 & 1896*.
87. Quoted in Gesink, *Islamic Reform and Conservatism*, 145.
88. "Hadithat al-Azhar," *al-Muqtataf* 20, no. 7 (1896): 558–59; Ministry of Interior, Sanitary Department, *Report on the Epidemic of Cholera in Egypt During the Years 1895 & 1896*, 97–98.
89. Ministry of Interior, Sanitary Department, *Report on the Epidemic of Cholera in Egypt During the Years 1895 & 1896*, 58.
90. Ibid.
91. Ibid., 97.
92. White, *Speaking with Vampires*, ch. 2.
93. TNA, FO 407/119, Lord Cromer to the Earl of Rosebery, March 9 (1893).

Chapter 4

1. Darwish, *Badi' Khayri*; Bahjat, *Badi' Khayri*.
2. Bahjat, *Badi' Khayri*, Part II, 343.

يهون الله يعوض الله ع السقايين دول شقيانين متعفرتين م الكوبانية
خواجاتها جونا حايطفشونا ليه بيرازونا دي صنعة أبونا ما تعبرونا يا خلايق
دي شركة غلسه ميتها نجسه تلقاها حدقه وخضره وزرقه وقال إيه رايقه دي لهجة
إخيه عليها بيحطوا فيها كربونات ونبيت وفوسفات كبريت وبودرة عفريت ها أو أو

3. Owen, *Cotton and the Egyptian Economy*, ch. 10; Owen, *The Middle East in the World Economy*, ch. 9.
4. Owen, "The Cairo Building Industry and the Building Boom of 1897 to 1907."
5. Jakes, *Egypt's Occupation*.
6. Abu Lughod, *Cairo*; Arnaud, *Le Caire*; Volait, "Making Cairo Modern."
7. "Rapport de la commission internationale de l'assainissement du Caire," *Journal Officiel*, supplement to no. 51 (April 16, 1892), 6; DWQ, Majlis al-Wuzara' (Council of Ministers), file no. 0075-033264, "Cairo Main Drainage: Preliminary Report," by Charles Carkeet James, October 22 (1907).

8. Foucart, "Le Caire et la spéculation," 488.

9. Artin, *Essai sur les causes du renchérissement de la vie matérielle au Caire*, 115.

10. Eid, *La Fortune Immobilière de l'Égypte*, 54.

11. Ahmad Hilmi, "'Umran al-'Asima," *al-Liwa'*, April 28, 1903.

12. Taylor *Egyptian Company Manual*, 142.

13. Krämer, *The Jews in Modern Egypt*, 39–41; Vitalis, *When Capitalists Collide*, 33–38.

14. Artin, "Le Premier Puits Artésien dans la Vallée du Nil."

15. Ministère de L'Intérieur, Département de l'Hygiène Publique, *Les Eaux d'Égypte* by A. Azadian, 20.

16. Artin, "Le Premier Puits Artésien dans la Vallée du Nil," 30.

17. Bitter, *Observations hygiéniques*, 19–23.

18. Artin, "Le Premier Puits Artésien dans la Vallée du Nil," 33.

19. Artin, *De l'eau du nil et de l'eau souterraine*, 16. For Arabic translation, see Yacoub Artin, "Ma' al-Nil wa al-Miyah al-Batiniya I," *Tabib al-'A'ila* 2 (November, 1896): 3–7, and idem, "Ma' al-Nil wa al-Miyah al-Batiniya II," *Tabib al-'A'ila* 2 (December, 1896): 24–29.

20. Guerville, *La nouvelle Égypte*, 34–36.

21. "Cairo Water Supply—Official Report (*suite*)," *The Egyptian Gazette*, July 18, 1906; See also Ministry of Finance, Survey Department, *The Chemistry of River Nile*, 40.

22. "Egypt (From our own Correspondent)," *The Lancet*, 9 July 1892, 120; Wright and Cartwright, *Twentieth Century Impressions of Egypt*, 417.

23. Bitter, *Observations hygiéniques.*

24. Ibid., 7–11.

25. Ibid., 29–30.

26. Willcocks, *Sixty Years in the East*, 169. See also Parodi, "La fabrication des engrais au Caire," 427.

27. Willcocks, *Sixty Years in the East*, 170–71.

28. DWQ, Diwan al-Ashghal al-'Umumiyya (Public Works Ministry), file no. 4003-037343, Draft project by William Willcocks, manager of the Cairo Water Company, June (1898); DWQ, Majlis al-Wuzara' (Council of Ministers), file no. 0075-033276, From the Council of Ministers to the international sanitary commission concerning Cairo drainage project by William Willcocks, July (1898).

29. Crouchley, *The Investment of Foreign Capital in Egyptian Companies*, 52.

30. Ibid., 52–53.

31. Quoted in Noyes, "A Year After the Panic of 1907," 203.

32. See classic studies such as: Hobson, *Imperialism*; Luxemburg, *The Accumulation of Capital*; Lenin, *Imperialism.*

33. Jakes, *Egypt's Occupation*, ch. 3

34. Kindleberger and Aliber, *Manias, Panics, and Crashes*, 19.

35. Galbraith, *A Short History of Financial Euphoria*; Quinn and Turner, *Boom and Bust*.

36. Chalcraft, *The Striking Cabbies of Cairo*, 132–41. See also Saadeh, "Motor Progress in the Land of the Pyramids," *The Horseless Age*, August 21, 1907, 239–41.

37. Boulad, *Les Tramways du Caire*; Clerget, *Le Caire*, Vol. 2, 107–8; Abu Lughod, *Cairo*, 279–85.

38. Kilani, *Tiram al-Qahira*.

39. DWQ, Majlis al-Wuzara' (Council of Ministers), file no. 0075-008608. Memorandum by the Public Works Ministry, January 16 (1897).

40. TNA, FO 371/248, Parliamentary Question from Mr. MacNeill, June 18 (1907).

41. "Richer Egypt, Magic of English Security, A Land Boom, The Potent Harem, Villa and Flat in Cairo," *The Egyptian Gazette*, July 2, 1906. See descriptions of Cairo's poor housing in Eid, *La Fortune immobilière de l'Égypte*, 61 and Corthell, "A Recent Trip to Italy, Egypt, Southern France and Spain," 263.

42. "Richer Egypt, Magic of English Security, A Land Boom, The Potent Harem, Villa and Flat in Cairo," *The Egyptian Gazette*, July 2, 1906.

43. Foucart, "Le Caire et la spéculation," 489.

44. Frapier and Vaillant, "The organization of the Hennebique firm in the countries of the Mediterranean Basin."

45. Kupferschmidt, "Who Needed Department Stores in Egypt?"; Reynolds, *A City Consumed*, ch. 2

46. Rafaat, *Maadi*, 11–32.

47. Ilbert, *Héliopolis*.

48. DWQ, Nizarat al-Dakhiliyya (Ministry of Interior), files no. 2001-018153 and 2001-025493, Reports and correspondences on typhoid fever in 'Abbasiyya, March and April (1889).

49. Rafaat, *Maadi*, 22–23.

50. Ilbert, *Héliopolis*, 50, 104, 115–17. See also Vallet, *Contribution à l'étude de la condition des ouvriers de la grande industrie au Caire*, 66–74.

51. Wright and Cartwright, *Twentieth Century Impressions of Egypt*, 362.

52. Ibid., 365–66.

53. DWQ, Majlis al-Wuzara' (Council of Ministers), file no. 0075-008614, Contract between the Egyptian Government and the Société Générale Égyptienne des Engrais (General Egyptian Fertilizers Company), March (1889); Cairo Sewage Transport Company advertisement in *al-Mu'ayyad*, February 2, 1900; Parodi, "La fabrication des engrais au Caire," 425–28; Wright and Cartwright, *Twentieth Century Impressions of Egypt*, 354.

54. Arminjon, "La Crise Financière Égyptienne Actuelle," 204, 211.

55. "Land Boom in Egypt. Speculative Builder in Cairo. Desert Suburbs," *The Egyptian Gazette*, July 4, 1906.

56. Arminjon, "La Crise Financière Égyptienne Actuelle," 204.

57. Galbraith, *A Short History of Financial Euphoria*, 3.

58. Kaufmann, "Sur le prétendu Nil vert."

59. *Convention pour l'alimentation de la ville du Caire par l'eau de puits artésiens* [Convention for the supply of the city of Cairo with water from artesian wells] January 25, 1903, in Ministère des Finances, *Recueil Général de Contrats* (Le Caire: Imprimerie Nationale, 1908), Preamble.

60. Ibid., Article 2.

61. *Procès-verbal de l'assemblée générale extraordinaire de la Société Anonyme des Eaux du Caire approvant la convention passée le 25 Avril 1903* [Minutes of the extraordinary general meeting of the Cairo Water Company approving the agreement signed on April 25, 1903], April 25, 1903, in Ministère des Finances, *Recueil Général de Contrats.*

62. "Cairo Water Supply—Official Report," *The Egyptian Gazette*, July 17, 1906; "Cairo Water Supply—Official Report (suite)," *The Egyptian Gazette*, July 18, 1906.

63. Wright and Cartwright, *Twentieth Century Impressions of Egypt*, 340.

64. *Convention pour la vente des terrains de la Salpêtrière* [Convention for the Sale of Land of the Saltpeter Works], March 7, 1892, in Ministère des Finances, *Recueil Général de Contrats*, 127–28.

65. Public Works Ministry, *Report Upon the Administration of the Public Works Department in Egypt for 1904*, 283.

66. Taylor, *Egyptian Company Manual*, 143.

67. Ibid.

68. For more on Zaydan and the Arab intellectual renaissance, see Elshakry, *Reading Darwin in Arabic*, ch. 1.

69. "Al-Nahda al-Maliyya al-Misriyya: Wahmiyya hiya am Haqiqiyya? [Part I]" *al-Hilal* 14 (May, 1906): 453.

70. Ibid., 460.

71. "Al-Nahda al-Maliyya al-Misriyya: Wahmiyya hiya am Haqiqiyya? [Part II]" *al-Hilal* 14 (June, 1906): 544.

72. "Al-Ghala' aw Irtifa' al-As'ar," *al-Hilal* 15 (May, 1907): 471–81.

73. "Al-Ihsa' wa al-Nahda al-Maliyya," *al-Hilal* 15 (June, 1907): 538.

74. Egypt, Ministry of Finance, *Ti'dad Sukkan al-Qutr al-Misri fi Sanat 1325 Hijriyya-1907 Miladiyya*, 25.

75. Artin, *Essai sur les causes du renchérissement de la vie matérielle au Caire*, 92–93.

76. For more on market inspectors (sing. *muhtasib*), see Fahmy, *In Quest of Justice*, ch. 4.

77. Artin, *Essai sur les causes du renchérissement de la vie matérielle au Caire*, 102.

78. Ibid., 112.

79. Ibid.

80. Wright and Cartwright, *Twentieth Century Impressions of Egypt*, 116.

81. Eid, *La Fortune Immobilière de l'Égypte*, 54, 66–71.

82. Arminjon, "La Crise Financière Égyptienne Actuelle," 215.

83. Ibid.

84. Ibid., 216

85. For a political biography of Hilmi see al-Musallami, *Ahmad Hilmi.*

86. Jakes, *Egypt's Occupation*, 113–27.

87. Himli, "'Umran al-'Asima."

88. Quoted in "The Earl of Cromer's Report on Egypt for 1905," *The Lancet*, May 12, 1906, 1339.

89. "Egypt (From our own Correspondent)," *The Lancet*, June 9, 1906, 1648.

90. DWQ, Majlis al-Wuzara' (Council of Ministers), file no. 0075-016454, Requests by the General Assembly concerning the bad quality of artesian well water (1907), file no. 0075-033494, From the Council of Ministers to the General Assembly (1907); Sharubim, *Al-Kafi*, Part 5, Vol. 3, 435–36.

91. TNA, FO 371/248, From Lord Cromer to Sir Edward Grey, April 2 (1907).

92. [Adolphe Smith], "A Crisis in the Water-Supply of Cairo (From Our Special Sanitary Commissioner)," *The Lancet*, April 3, 1909, 1014.

93. "Cairo Water Supply—Official Report," *The Egyptian Gazette*, July 17, 1906.

94. "Cairo Water Supply—Official Report (*suite*)," *The Egyptian Gazette*, July 18, 1906.

95. Bitter, *Observations hygiéniques*, 30.

96. "Egypt (From our own Correspondent)," *The British Medical Journal*, June 9, 1906, 1648.

97. TNA, FO 371/662, Cairo Water Supply, July 12 (1909).

98. Al-Rafi'i, *Mustafa Kamil*, ch. 12.

99. TNA, FO 407/172, Sir E. Gorst to Sir Edward Grey, dated 7 March (1908).

100. See also Ismail, "Epicures and Experts."

101. "Cairo Water Supply—Official Report (suite)," *The Egyptian Gazette*, July 18, 1906 and "Cairo Water Supply," *The Egyptian Gazette*, July 24, 1906.

102. Ministry of Interior, Department of Public Health, *Annual Report for 1909*, 72–78.

103. Gilberg, "Alfred Lucas."

104. Ministry of Finance, Survey Department, *The Chemistry of River Nile*, 40–41.

105. Smith, "The Subsoil of Cairo."

106. [Adolphe Smith], "A Crisis in the Water-Supply of Cairo (From Our Special Sanitary Commissioner)," *The Lancet*, April 3, 1909, 1014.

107. Ibid., 1016.

108. Ministry of Interior, Department of Public Health, *Annual Report for 1909*, 72–73.

109. Bitter and Blagden, *Filtration Experiments at Alexandria*; "Cairo's Water Supply—The Battle of the Filters—Attitude of Ministers," *The Egyptian Gazette*, February 19, 1910; Cottarel, *Bahth Tarshih Miyah al-Nil al-Mu'adda li al-Shurb.*

110. Noyes, "A Year After the Panic of 1907," 206.

111. Jakes, *Egypt's Occupation*, 143–49.

112. Foucart, "Le Caire et la spéculation," 488.

113. Quoted in Ilbert, *Héliopolis*, 17.

114. Sladen, *Oriental Cairo*, 37.

115. Foucart, "Le Caire et la spéculation," 489.

116. Wasfi, *Al-Qawa'id al-Asasiyya*, Vol. 2, 7.

117. Ahmad Lutfi al-Sayyid, "Al-Sihha al-'Amma wa Mustaqbal al-Umma," *al-Jarida*, December 8, 1909.

118. Ibid.

119. James, "The Main Drainage of Cairo," 61.

120. Ibid.

121. [Adolphe Smith], "The Drainage of Cairo (From Our Special Sanitary Commissioner)," *The Lancet*, April 24, 1909, 1210.

122. Ibid.

123. Jacob, *Working Out Egypt*; Reynolds, *A City Consumed*; Ryzova, *The Age of the Efendiyya*.

Chapter 5

1. Home, *Of Planting and Planning*, 167.

2. FO 141/581, From William McLean to Arthur Henderson, October 3 (1925).

3. Cherry, "The Town Planning Movement in the Late Victorian City"; Schultz and McShane, "To Engineer the Metropolis"; Peterson, "The Impact of Sanitary Reform upon American Urban Planning"; Claude, "Sanitary Engineering as a Path to Town Planning."

4. Tyrwhitt *Patrick Geddes in India*, 34–83; Home, *Of Planting and Planning*, 175–80.

5. Meller, *Patrick Geddes*; Batty and Marshall, "The Evolution of Cities"; Clavel and Young, ""Civics"."

6. FO 407/172, Note by the Financial Adviser on the Budget of the Egyptian Government for the year 1908 (1908); James, "The Main Drainage of Cairo," 95.

7. FO 407/172, From Sir E. Gorst to Sir Edward Grey, March 7 (1908); James, "The Main Drainage of Cairo," 94.

8. DWQ, Majlis al-Wuzara' (Council of Ministers), file no. 0075-034678, Memorandum by the Public Works Ministry, March 29 (1906).

9. Institution of Civil Engineers Archives, United Kingdom, Charles Carkeet James papers.

10. James, *Drainage Problems of the East*.

11. DWQ, Majlis al-Wuzara' (Council of Ministers), file no. 0075-033279, "Precis of Each of the Proposed Schemes for the Sewerage of Cairo," February 22 (1908).

12. DWQ, Majlis al-Wuzara' (Council of Ministers), file no. 0075-033280, Note by William Garstin on the preliminary report by Charles Carkeet James, February 22 (1908).

13. Public Works Ministry, *Report Upon the Administration of the Public Works Department for 1898*, 35–38, 205–17.

14. DWQ, Majlis al-Wuzara' (Council of Ministers), file no. 0075-033280, Note by William Garstin on the preliminary report by Charles Carkeet James, February 22 (1908).

15. DWQ, Majlis al-Wuzara' (Council of Ministers), file no. 0075-033264, "Cairo Main Drainage: Preliminary Report," by Charles Carkeet James, October 22 (1907).

16. Public Works Ministry, *Report upon the Administration of Public Works Department in Egypt for 1904*, 283.

17. DWQ, Majlis al-Wuzara' (Council of Ministers), file no. 0075-033264, "Cairo Main Drainage: Preliminary Report," by Charles Carkeet James, October 22 (1907); Ministry of Public Works, *Sewerage of Cairo, First Section, Report and Drawings*, 13.

18. DWQ, Majlis al-Wuzara' (Council of Ministers), file no. 0075-033264, "Cairo Main Drainage: Preliminary Report," by Charles Carkeet James, October 22 (1907).

19. DWQ, Majlis al-Wuzara (Council of Ministers), file no. 0075-033264, "Cairo Main Drainage: Preliminary Report," by Charles Carkeet James, October 22, 1907. See also James, "The Main Drainage of Cairo," 63.

20. TNA, FO 407/170, The Earl of Cromer to Sir Edward Grey, March 3 (1907).

21. Maunier, "La Vie Municipale en Egypte"; McLean, "Local Government and Town Development in Egypt"; Shalabi, *Al-Hukm al-Mahalli wa al-Majalis al-Baladiyya fi Misr.*

22. Reimer, "Urban Government and Administration in Egypt," 295.

23. Al-Rafi'i, *Mustafa Kamil.*

24. For more on constitutional revolutions in the Middle East, see Sohrabi, *Revolution and Constitutionalism.*

25. Jakes, *Egypt's Occupation*, ch. 6

26. Ahmad Hilmi, "'Umran al-'Asima," *al-Liwa'*, April 28, 1903.

27. Al-Jam'iyya al-'Umumiyya, *Mahadir Jalasat al-Jam'iyya al-'Umumiyya*, Vol. 1, 326.

28. Caneri, *La Ville du Caire.*

29. Al-Jam'iyya al-'Umumiyya, *Mahadir Jalasat al-Jam'iyya al-'Umumiyya*, Vol. 1, 417–20; 493–94.

30. Ibid., Vol. 1, 390.

31. TNA, FO 407/170, From British Chamber of Commerce, Egypt, to the Earl of Cromer, March 9, 1907; From Cairo Financiers to the Earl of Cromer, March 6 (1907).

32. TNA, FO 407/174, From Sir E. Gorst to Edward Grey, March 27 (1909).

33. Ibid.

34. TNA, FO 371/662, Report of the Special Study Committee, May 31 (1909).

35. *Al-Mu'ayyad*, February 1, 1909.

36. Al-Jam'iyya al-'Umumiyya, *Mahadir Jalasat al-Jam'iyya al-'Umumiyya*, Vol. 1, 491.

37. Al-Rafi'i, *Muhammad Farid*, 160–61.

38. Amin al-Rafi'i was the elder brother of 'Abd al-Rahman al-Rafi'i—widely considered Egypt's foremost nationalist historian. See 'Anbar, *Dhikra Faqid al-Watan al-Maghfur Lahu Amin al-Rafi'i*, 16–17.

39. Al-Jam'iyya al-'Umumiyya, *Mahadir Jalasat al-Jam'iyya al-'Umumiyya*, Vol. 2, 403.

40. Seikaly, "Prime Minister and Assassin," 121; Badrawi, *Political Violence in Egypt*, ch. 3; Khuri-Maqdisi, *The Eastern Mediterranean*, 129–31.

41. FO 407/175, Memorandum by Dr. Nolan on the Wardani Case, February 20 (1910); Badrawi, *Political Violence in Egypt*, ch. 3.

42. Moore, *Images of Development*, 24; Derr, *The Lived Nile*, 56.

43. "Bi'ayy Qanun," *al-Sha'b*, April 12, 1910.

44. Jamil 'Arif, "Limadha Haribt min Baladi!" *Akhir Sa'a*, March 16, 1960.

45. Ministry of Public Works, *Sewerage of Cairo, First Section, Report and Drawings*; Ministry of Public Works, *Sewerage of Cairo, Report on the Second Section*; Egyptian Government, Ministry of Public Works, Main Drainage Department, *General Conditions as to the Execution of Works Under Contract with the Main Drainage Department.*

46. Egyptian Government, Ministry of Public Works, Main Drainage Department, *Contract No. 1.*

47. Egyptian Government, Ministry of Public Works, *Report of the Ministry of Public Works for the Year 1911*, 478.

48. "Majari al-'Asima," *al-Sha'b*, April 27, 1910; "Majari al-'Asima," *al-Sha'b*, May 12, 1910; "Majari al-'Asima," *al-Sha'b*, May 17, 1910; "Majari al-'Asima," *al-'Alam*, May 30, 1910.

49. "Bi'ayy Qanun," *al-Sha'b*, April 12, 1910.

50. "Majari al-'Asima," *al-Sha'b*, May 12, 1910.

51. "Majlis Shura al-Qawanin, Jalsat Ams," *al-Sha'b*, May 6, 1910.

52. "Majari al-'Asima," *al-'Alam*, May 30, 1910.

53. Sabett [*sic*], "L'hygiène publique en Égypte."

54. Smith was the author of *Street Life in London* (1876–77)—a series of articles focused on the city's poor and working class. See Morgan, "'True Types of the London Poor.'"

55. [Adolphe Smith], "The Drainage of Cairo (From Our Special Sanitary Commissioner)," *The Lancet*, May 8, 1909.

56. Ibid., 1353.

57. DWQ, Majlis al-Wuzara' (Council of Ministers), file no. 0075-035852, Note from the Public Works Ministry to the Council of Ministers, August 23 (1911).

58. Egyptian Government, Ministry of Public Works, *Report of the Ministry of Public Works for the Year 1911*, 473–77.

59. Hodge, *Triumph of the Expert*, 64–70.

60. Smith, "The Cairo Sewage Farm at Gebel el Asfar," 24.

61. Egyptian Government, Ministry of Public Works, *Report of the Ministry of Public Works for the Year 1911*, 475; James, "The Main Drainage of Cairo," 91.

62. Correspondence on James, "The Main Drainage of Cairo," 128.

63. Ibid.

64. Correspondence on James, "The Main Drainage of Cairo," 135.

65. Ibid., 135–36.

66. Tyrwhitt, *Patrick Geddes in India*, 45.

67. Ibid., 41.

68. Egyptian Government, Ministry of Public Works, *Report of the Ministry of Public Works for the Year 1911*, 355.

69. Ibid.

70. Ibid., 355–56. This mechanism is now referred to as "land value capture." For more information, see OECD/Lincoln Institute of Land Policy, PKU-Lincoln Institute Center (2022), *Global Compendium of Land Value Capture Policies*.

71. DWQ, Majlis al-Wuzara' (Council of Ministers), file no. 0075-035852, Note from the Public Works Ministry to the Council of Ministers, August 23 (1911).

72. FO 141/633, Note by Murdoch Macdonald on the Tanzim Department, May 23 (1915).

73. Egyptian Government, Ministry of Public Works, *Report of the Ministry of Public Works for the Year 1914–1915*, 68.

74. Home, "British Colonial Town Planning in the Middle East."

75. McLean, "Local Government and Town Development in Egypt."

76. FO 141/581, From William McLean to Arthur Henderson, October 3 (1925).

77. Egypt, Ministry of Public Works, *Annual Report: 1915–1916*, 14.

78. Ibid.

79. Ibid., 12.

80. Egypt, Ministry of Public Works, *Annual Report for 1918–1919*, 48.

81. Ibid., 47.

82. Egypt, Ministry of Public Works, *Annual Report for 1920–1921*, 5, 116.

83. Haswell, "Town Planning and Housing in Cairo."

84. Ibid., 258.

85. Municipality of Alexandria, *City of Alexandria Town Planning Scheme* by W. H. McLean.

86. TNA, FO 141/581, Note on the Necessity for a General Project of Development, National, Regional, and Urban in Egypt by William McLean, October 10 (1923).

87. Havinden and Meredith, *Colonialism and Development*; Cooper and Packard, *International Development and the Social Sciences*; Drayton, *Nature's Government*; Hodge, *Triumph of the Expert.*

88. Home, "British Colonial Town Planning in the Middle East." Henry Lanchester, British urban planner and an associate of Geddes, heavily criticized McLean's plan for Jerusalem: "We cannot but accuse (Mr. McClean) [*sic*] of being, to say the least, somewhat disrespectful towards his art as a town planner in submitting a scheme for the laying out of the selected area so lacking in even superficial study of the site and conditions." Quoted in Meller, *Patrick Geddes*, 196.

89. Jam'iyyat al-Muhandisin al-Malakiyya al-Misriyya, *Kitab Jam'iyyat al-Muhandisin al-Malakiyya al-Misriyya* 1, no. 1.

90. Moore, *Images of Development*, 27–28.

91. For a study of Sudan as an Egyptian colony see Powell, *A Different Shade of Colonialism.*

92. Nile Projects Commission, *Minutes of the Discussion Between the Nile Projects Commission and the Committee of Egyptian Engineers.*

93. A letter from William Willcocks to Prime Minister 'Adli Yakan (1921), cited in Zaki, *Isti'bad Misr*, 25; Derr, *The Lived Nile*, 62–63.

94. Zaki, *Isti'bad Misr*, 7.

95. Moore, *Images of Development*, 27.

96. Ibid., 28.

97. TNA, FO 141/633, Cairo Tanzim Department, Annual Report 1923–24, August 1 (1924).

98. On the tramway strike, see Beinin and Lockman, *Workers on the Nile*, 91–95.

99. TNA, FO 141/633, Cairo Tanzim Department, Annual Report 1923–24, August 1 (1924).

100. TNA, FO 141/633, Note on the Tanzim Department by A. Mann Alcock, January 4 (1925).

101. Ibid.

102. Ibid.

103. Fahmi, *Safwat al-'Asr*, 355–63.

104. Volait, "Town Planning Schemes for Cairo Conceived by Egyptian Planners in the "Liberal Experiment" Period."

105. Mahboub, "Cairo."

Chapter 6

1. Darwish, *Badi' Khayri*; Bahjat, *Badi' Khayri.*

2. Beinin and Lockman, *Workers on the Nile*, 91–95.

3. Bahjat, *Badi' Khayri*, Part II, 452.

تشوفنا إحنا تقول حنفيه وتشوفه هوه تقول بلاعه
عايشين في وادي النيل نشرب بالعدادات على مللي وسنتي

4. Majlis al-Nuwwab, *Madabit Majlis al-Nuwwab*, session of April 24, 1934.

5. Veblen, *The Engineers and the Price System*.

6. *Tarif Général pour la ville du Caire* [General Tariff for the City of Cairo], January 1, 1887, in Ministère des Finances, *Recueil Général de Contrats*.

7. American Water Works Association, *Water Meters*, ch.1.

8. DWQ, Majlis al-Wuzara' (Council of Ministers), file no. 0075-059294, Explanatory note regarding the 1938 agreement between the government and the water company (1941).

9. Reyburn, *Flushed with Pride*, ch. 3.

10. DWQ, Majlis al-Wuzara' (Council of Ministers), file no. 0075-059294, Explanatory note regarding the 1938 agreement between the government and the water company (1941).

11. Ministry of Public Works, *Annual Report for 1916–1917*, 19–20; Ministry of Public Works, *Annual Report for 1919–1920*, 120–21.

12. DWQ, Majlis al-Wuzara' (Council of Ministers), file no. 0075-059294, Explanatory note regarding the 1938 agreement between the government and the water company (1941).

13. Ibid.

14. DWQ, Majlis al-Wuzara' (Council of Ministers), file no. 0075-059295, "Report to his Excellency the Minister of Public Works on the charges for water in Cairo," by S. R. Raffety, February 4 (1938).

15. Wizarat al-Ashghal al-'Umumiyya, *Al-Taqrir al-Sanawi li Sanat 1925–1926*, 122; Wizarat al-Ashghal al-'Umumiyya, *Al-Taqrir al-Sanawi li Sanat 1926–1927*, 163; Wizarat al-Ashghal al-'Umumiyya, *Al-Taqrir al-Sanawi li Sanat 1927–1928*, 168.

16. Egyptian Government, Ministry of Public Works, *Annual Report for 1920–1921*, 126.

17. Egyptian Government Press, *Almanac for the Year 1929*, 246.

18. Egyptian Government Press, *Almanac for the Year 1933*, 309.

19. Ibid., 308.

20. Ibid., 310.

21. Ibid., 308.

22. Ibid., 309.

23. Ibid., 304.

24. Wasfi, *Muhadara 'an Majari al-Qahira*, 27.

25. Ministry of Public Works, *Annual Report for 1916–1917*, 19; Egyptian Government Press, *Almanac for the Year 1933*, 305.

26. Egyptian Government Press, *Almanac for the Year 1933*, 305.

27. Allen, "Cairo Drainage Scheme."

28. *'Aridat al-Da'wa*, December 19, 1932, in Nazmi, *Bahth Wajiz*, 30–44.

29. Ibid., 31.

30. Majlis al-Nuwwab, *Madabit Majlis al-Nuwwab*, session of April 24, 1934.

31. For instance, Nazmi participated in the Higher Schools Club (Nadi al-Madaris al-ʻUlya) and events organized by the Egyptian Feminist Union. See Al-Rafiʻi, *Mustafa Kamil*, 197 and Shaʻrawi, *Mudhakirat Huda Shaʻrawi*, 204.

32. Majlis al-Nuwwab, *Madabit Majlis al-Nuwwab*, session of April 24, 1934.

33. Ibid.

34. Majlis al-Nuwwab, *Madabit Majlis al-Nuwwab*, session of April 25, 1934.

35. Majlis al-Nuwwab, *Madabit Majlis al-Nuwwab*, session of April 24, 1934.

36. Ibid.

37. Cited in *ʻAridat al-Daʻwa*, in Nazmi, *Bahth Wajiz*, 39.

38. *Procès-verbal de l'assemblée générale extraordinaire de la Société Anonyme des Eaux du Caire approvant la convention du 27 Janvier 1887* [Minutes of the Extraordinary General Meeting of the Cairo Water Company Approving the Convention of January 27, 1887], February 22, 1887, in Ministère des Finances, *Recueil Général de Contrats.*

39. Majlis al-Nuwwab, *Madabit Majlis al-Nuwwab*, session of April 24, 1934.

40. Quoted in*ʻAridat al-Daʻwa*, in Nazmi, *Bahth Wajiz*, 40.

41. Majlis al-Nuwwab, *Madabit Majlis al-Nuwwab*, session of April 25, 1934.

42. Ibid.

43. *Al-Muqattam*, April 27, 1934.

44. *Al-Ahram*, April 25, 1934.

45. *Al-Shaʻb*, April 27, 1934.

46. *Al-Musawwar*, May 4, 1934.

47. Goldschmidt, *Biographical Dictionary*, 188–89.

48. DWQ, Majlis al-Wuzara' (Council of Ministers), file no. 0075-059294, Explanatory note regarding the 1938 agreement between the government and the water company (1941).

49. Ibid.

50. Ibid.

51. "Obituary: Sidney Robert Raffety, CBE, 1883–1965," *Proceedings of the Institution of Civil Engineers* 35, no. 2 (1966): 384.

52. DWQ, Majlis al-Wuzara' (Council of Ministers), file no. 0075-059295, "Report to his Excellency the Minister of Public Works on the Charges for Water in Cairo," by S. R. Raffety, February 4 (1938).

53. Dobb, *Theories of Value and Distribution*, 43–58, 122–31.

54. Perelman, *Railroading Economics*, 40–42; Kurz, *Economic Thought*, ch. 2.

55. Henderson, "Walter Rathenau"; Layton, *The Revolt of the Engineers*, chs. 5–8; Stabile, "Veblen and the Political Economy of the Engineer"; Knoedler and Mayhew, "Thorstein Veblen and the Engineers."

56. Veblen, *The Engineers and the Price System*. See also: Veblen, *The Theory of Business Enterprise*; Baran and Sweezy, *Monopoly Capital*, 28–29, 132; Nitzan,

"Differential Accumulation"; Hunt and Lautzenheiser, *History of Economic Thought*, ch. 12.

57. DWQ, Majlis al-Wuzara' (Council of Ministers), file no. 0075-059295, "Report to his Excellency the Minister of Public Works on the Charges for Water in Cairo," by S. R. Raffety, February 4 (1938).

58. Ibid.

59. Ibid.

60. Shadwell, *The London Water Supply*, 264–68.

61. DWQ, Majlis al-Wuzara' (Council of Ministers), file no. 0075-059295, "Report to his Excellency the Minister of Public Works on the Charges for Water in Cairo," by S. R. Raffety, February 4 (1938).

62. See also Issawi and Rosenfeld, "Company Profits in Egypt," 686.

63. DWQ, Majlis al-Wuzara' (Council of Ministers), file no. 0075-059295, "Report to his Excellency the Minister of Public Works on the Charges for Water in Cairo," by S. R. Raffety, February 4 (1938).

64. Ibid.

65. Ibid.

66. Ibid.

67. Ibid.

68. Ibid.

69. Ibid.

70. Ibid.

71. DWQ, Majlis al-Wuzara' (Council of Ministers), file no. 0075-059294, Explanatory note regarding the 1938 agreement between the government and the water company (1941). See also Chehata, "La Concession de Service Public," 364–68.

72. "Law no. 127 for the year 1946," *Al-Waqa'i' al-Misriyya*, September 2, 1946.

73. DWQ, Majlis al-Wuzara' (Council of Ministers), file no. 0075-059294, Explanatory note regarding the 1938 agreement between the government and the water company (1941).

74. Rizq, *Tarikh al-Wizarat al-Misriyya*, 387–431; Deeb, *Party Politics in Egypt*; Beinin and Lockman, *Workers on the Nile*, chs. 5–8.

75. Goldschmidt, *Biographical Dictionary of Modern Egypt*, 116–17.

76. Al-Rafi'i, *Thawrat 1919*, 245–46.

77. Majlis al-Nuwwab, *Madabit Majlis al-Nuwwab*, session of July 13, 1938.

78. Ibid.

79. Ibid.

80. Ibid.

81. Ibid.

82. Ibid.

83. Ibid.

84. Ibid.

85. 'Alluba, *Mabadi' fi al-Siyasa al-Misriyya*, 86–88.

86. *Al-Musawwar*, July 29, 1938.

87. *Al-Musawwar*, July 29, 1938.

88. Majlis al-Shuyukh, *Madabit Majlis al-Shuyukh*, session of February 3, 1941.

89. Ibid.

90. DWQ, Majlis al-Wuzara' (Council of Ministers), file no. 0075-059294, Explanatory note regarding the 1938 agreement between the government and the water company (1941).

91. "Law no. 127 for the year 1946," *Al-Waqa'i' al-Misriyya*, September 2, 1946.

92. Beinin and Lockman, *Workers on the Nile*, 257–58.

93. Vitalis and Heydemann, "War, Keynesianism, and Colonialism."

94. Beinin and Lockman, *Workers on the* Nile, chs. 8–11; Meijer, *The Quest for Modernity*.

95. Al-Shafi'i, *Tatawwur al-Haraka al-Wataniyya al-Misriyya*, 95–109; Al-Bishri, *Al-Haraka al-Siyasiyya fi Misr*, chs. 4–5.

96. Al-Siba'i, "Sadiq Sa'd"; Beinin and Lockman, *Workers on the Nile*, 314–21.

97. Sa'd, *Ma'sat al-Tamwin*.

98. Sa'd, *Mushkilat al-Fallah*; See also: Abdel-Malek, *Egypt, Military Society*, 62–65; Shokr, *Harvests of Liberation*, ch. 5.

99. Shakry, "Cairo as Capital of Socialist Revolution?"

100. Madani, *Mas'uliyyat al-Dawla 'an A'amaliha al-Musharra'a*, 375–92.

101. Ibid., 382.

102. "Law no. 129 for the year 1947," *Al-Waqa'i' al-Misriyya*, July 24, 1947.

103. Madani, *Mas'uliyyat al-Dawla 'an A'amaliha al-Musharra'a*, 382.

104. "Law no. 138 for the year 1947," *Al-Waqa'i' al-Misriyya*, August 4, 1947.

105. The government's attempt to apply this law to the Suez Canal Company sparked a legal dispute. See Madani, *Mas'uliyyat al-Dawla 'an A'amaliha al-Musharra'a*, 378–84; Tignor, "Foreign Capital, Foreign Communities, and the Egyptian Revolution of 1952," 119–21.

106. "Law no. 145 for the year 1949," *Al-Waqa'i' al-Misriyya*, September 1, 1949. For more on Mahmud Riyad, see Elshahed, *Cairo since 1900*, 87–100, and Volait, "Garden Suburbs, Cairo 1940s–1950s Architect: Mahmoud Riad." On social housing projects during this period, see Shawkat, *Egypt's Housing Crisis*, 122–125.

107. Vitalis, *When Capitalists Collide*, 196–214.

108. Abdel-Malek, *Egypt, Military Society*, 88.

109. Quoted in Abdel-Malek, *Egypt, Military Society*, 89.

110. Abdel-Malek, *Egypt, Military Society*, 90.

111. Abul-Magd, *Militarizing the Nation*, ch. 1.

112. "Law no. 145 for the year 1957," *Al-Waqa'i' al-Misriyya*, July 1, 1957.

Conclusion

1. Graham and Marvin, *Splintering Urbanism.*

2. Tayia, Collins, and Gilmont, "The Role of Virtual-Water Decoupling in Achieving Food-Water Security."

3. Graham and Marvin, *Splintering Urbanism.*

4. "Oder no. 577 for the year 1957," *Al-Waqa'i' al-Misriyya,* July 1, 1957.

5. Samira Ghubryal, "Sirr Fayadanat al-Majari," *Al-Ahram,* October 25, 1963.

6. "Qabl an Yaghraq 170 Shari' fi al-Qahira Taht Miyah al-Tafh Marra Ukhra," *Al-Ahram,* March 29, 1969.

7. Waterbury, *The Egypt of Nasser and Sadat.*

8. "Law no. 203 for the year 1991," *Al-Jarida al-Rasmiyya,* June 19, 1991; "Order no. 135 for the year 2004," *Al-Jarida al-Rasmiyya,* April 29, 2004; Ismail, "Drinking Water Protests in Egypt"; ""Kubbayat al-Miyah" fi Marma Qanun Jadid: Hiwar ma'a 'Abd al-Mawla Isma'il," interview by Beesan Kassab, *Mada Misr,* July 15, 2025.

9. Ismail, "Drinking Water Protests in Egypt," 4.

10. Bayat and Denis, "Who is Afraid of *Ashwaiyyat*?"; Sims, *Understanding Cairo,* ch. 4; Shawkat, *Egypt's Housing Crisis,* ch. 2.

11. Mitchell, *Rule of Experts,* ch. 9; Sims, *Understanding Cairo,* ch. 5.

12. Sims, *Egypt's Desert Dreams.*

13. Denis, "Cairo as Neo-Liberal Capital?"

14. Quoted in Piper, *The Price of Thirst,* 164.

15. Ismail, "Drinking Water Protests in Egypt," 3.

16. Piper, *The Price of Thirst,* ch. 5.

17. Duffield, "Military Capitalism."

18. The total cost of Cairo's third metro line (finished in 2024) was approximately L.E. 97 billion, equivalent to $ 6 billion.

19. Khalil, "Cairo's Incarcerated Geographies."

20. Ahmad, *Ma'arik al-Miyah al-Muqbila fi al-Sharq al-Awsat,* 7–109; Wittington, Waterbury, and Jeuland, "The Grand Renaissance Dam."

21. "Why is Egypt Worried about Ethiopia's Dam on the Nile," *BBC News,* September 13, 2003. https://www.bbc.com/news/world-africa-66776733.

22. Faruq al-Baz counts as the country's foremost promoter of groundwater development.

23. Stone, *Radical Adaptation.*

24. Millington, "Producing Water Scarcity in São Paulo, Brazil"; Millington and Scheba, "Day Zero and The Infrastructures of Climate Change."

25. Farmer, "Willing to Pay."

26. Wahby, "Urban Informality and the State."

27. "Law no. 172 for the year 2025," *Al-Jarida al-Rasmiyya,* September 1, 2025.

28. McDonald, "Remunicipalization."

29. Le Strat, "Discussion: The Remunicipalization of Paris's Water Supply Service"; Valdovinos, "The Remunicipalization of Parisian Water Services."

30. Although, in July 2025, the Javier Milei government announced a plan to reprivatize 90 percent of state-run water and sanitation agencies.

31. "We Own it" campaign: https://weownit.org.uk/; George Monbiot, "High Water Bills, Filthy Rivers—and Now Drought. This is England's Great Artificial Water Crisis of 2025," *The Guardian*, July 17, 2025. https://www.theguardian.com/commentisfree/2025/jul/17/water-bills-filthy-rivers-drought-water-crisis.

BIBLIOGRAPHY

Archival Sources

American University in Cairo, Cairo, Rare Books and Special Collections Library
Photographic Collections
Map Collections

TNA: British National Archives, London
Foreign Office (FO): 141, 371, 407, 848

Centre d'études et de documentation économiques, juridiques et sociales (CEDEJ), Cairo
Map Collections

DWQ: Dar al-Watha'iq al-Qawmiyya (Egyptian National Archives), Cairo
Diwan al-Ashghal al-'Umumiyya (Ministry of Public Works) (4003)
Majlis al-Wuzara' (Council of Ministers) (0075)
Muhafazat Misr (Cairo Province) (2002)
Nizarat al-Dakhiliyya (Ministry of Interior) (2001)

Greater Cairo Sanitary Drainage Company, Cairo
Design Plans for Various Sewage System Projects (1915, 1929)

Al-Hay'a al-Misriyya al-'Amma li al-Misaha (Egyptian Survey Authority), Cairo
Map Collections

Institution of Civil Engineers Archives, London
Baldwin Latham Papers
Charles Carkeet James Papers

Selected Periodicals

Al-'Alam
Al-Azhar
The British Medical Journal
Bulletin de l'Institut Égyptien (Renamed *Bulletin de l'Institut d'Égypte*)
Cairo Scientific Journal
L'Égypte Contemporaine
The Egyptian Gazette
Al-Hilal
Journal Officiel
Kitab Jam'iyyat al-Muhandisin al-Malakiyya al-Misriyya
The Lancet
Al-Liwa'
Al-Majalla al-Tibbiyya al-Misriyya
Al-Mu'ayyad
Al-Muhandis
Al-Muqtataf
Al-Musawwar
Minutes of Proceedings of the Institution of Civil Engineers
Al-Sha'b
Al-Sihha
Tabib al-'A'ila
Al-Waqa'i' al-Misriyya

Government Sources

Egypt

Egyptian Government. *Qararat wa Manshurat*. Bulaq: Al-Matba'a al-Amiriyya, 1884.

Egyptian Government Press. *Almanac for the Year 1929*. Cairo: Government Press, 1929.

Egyptian Government Press. *Almanac for the Year 1933*. Cairo: Government Press,1933.

Al-Jam'iyya al-'Umumiyya (General Assembly). *Mahadir Jalasat al-Jam'iyya al-'Umumiyya*. 2 vols. Cairo: Dar al-Kutub wa al-Watha'iq al-Qawmiyya, 2011.

Majlis al-Nuwwab (House of Representatives). *Madabit Majlis al-Nuwwab*. 1934, 1938.

Majlis al-Shuyukh (House of Senate). *Madabit Majlis al-Shuyukh*. 1941.

Ministry of Finance. *Recueil Général de Contrats*. Le Caire: Imprimerie Nationale, 1908.

———. *Ti'dad Sukkan al-Qutr al-Misri fi Sanat 1325 Hijriyya-1907 Miladiyya*. Cairo: Bulaq Press, 1909.

———, Statistical Department. *Statistical Yearbook of Egypt for 1909*. Cairo: National Printing Department, 1909.

———, Statistical Department. *Report on Living Expenses* by T. L. Benet. Cairo: Government Press, 1921.

———, Survey Department. *The Chemistry of River Nile* by Alfred Lucas. Cairo: National Printing Department, 1908.

———, Survey Department. *The Cadastral Survey of Egypt, 1892-1907* by H. G. Lyons. Cairo: National Printing Department, 1908.

———, Survey of Egypt. *Geology of Egypt. Volume 1: The Surface Features of Egypt, their Determining Causes and Relation to Geological Structure* by W. F. Hume. Cairo, Government Press, 1925.

Ministry of Interior, Sanitary Department. *Report on the Epidemic of Cholera in Egypt During the Years 1895 & 1896*. Cairo: National Printing Office, 1897.

———, Department of Public Health. *Annual Report for 1909*. Cairo: National Printing Department, 1910.

———, Département de l'Hygiène Publique. *Les Eaux d'Égypte* by A. Azadian. Le Caire: Imprimerie Nationale, 1930.

Ministry of Public Works. *Résumé du rapport de M. Baldwin Latham sur le drainage du Caire*. Le Caire: Imprimerie Nationale, 1890.

———. *Report Upon the Administration of the Public Works Department for 1898*. Cairo: National Printing Office, 1899.

———. *Report Upon the Administration of the Public Works Department in Egypt for 1904*. Cairo: National Printing Office, 1905.

———. *Sewerage of Cairo, First Section, Report and Drawings*. Cairo: National Printing Department, 1909.

———, Main Drainage Department. *Contract No. 1: Conditions of Tender, Form of Tender, Special Conditions and Detailed Description of Work, and Schedule of Rates for Engineers' Quarters at Pont Sahel*. Cairo: National Printing Department, 1909.

———. *Sewerage of Cairo, Report on the Second Section*. Cairo: National Printing Department, 1910.

———. *Sewerage of Cairo, Photos of the Work in Progress, 1909-1910*. Cairo: National Printing Department, 1910.

———, Main Drainage Department. *General Conditions as to the Execution of Works Under Contract with the Main Drainage Department*. Cairo: Government Press, 1912.

———. *Report of the Ministry of Public Works for the Year 1911*. Cairo: Government Press, 1913.

———. *Report of the Ministry of Public Works for the Year 1914-1915*. Cairo: Government Press, 1916.

———. *Annual Report: 1915-1916*. Cairo: Government Press, 1917.

———. *Annual Report for 1916-1917*. Cairo: Government Press, 1921.

———. *Annual Report for 1918-1919*. Cairo: Government Press, 1922.
———. *Annual Report for 1919-1920*. Cairo: Government Press, 1923.
———. *Annual Report for 1920-1921*. Cairo: Government Press, 1923.
Wizarat al-Ashghal al-'Umumiyya [Ministry of Public Works]. *Al-Taqrir al-Sanawi li Sanat 1925-1926*. Cairo: Al-Matba'a al-Amiriyya, 1930.
———. *Al-Taqrir al-Sanawi li Sanat 1926-1927*. Cairo: Al-Matba'a al-Amiriyya, 1930.
———. *Al-Taqrir al-Sanawi li Sanat 1927-1928*. Cairo: Al-Matba'a al-Amiriyya, 1931.
Municipality of Alexandria. *City of Alexandria Town Planning Scheme* by W. H. McLean. Cairo: Government Press, 1921.

France

Commission des Sciences et des Arts. *Description De l'Égypte*, translated by Zuhayr al-Shayib and Muna Zuhayr al-Shayib. 37 vols. Cairo: Dar al-Shayib, 1992 [1809].
Ministère des Affaires Étrangères. *Documents Diplomatiques. Affaires d'Égypte, 1884-1893*. Paris: Imprimerie Nationale, 1893.

United Kingdom

British Foreign Office. *Correspondences respecting the cholera epidemic in Egypt: 1883*. Commercial, No. 34. London: Harrison and Sons, 1883.
———. *Further reports by surgeon-general Hunter on the cholera epidemic in Egypt*. Commercial, No. 38. London: Harrison and Sons, 1883.
———. *Further correspondences respecting the cholera epidemic in Egypt: 1883*. Commercial, No. 39. London: Harrison and Sons, 1883.

Other Published Sources

Abbate, Onofrio. "Questions hygiéniques sur la ville du Caire." *Bulletin de l'Institut Égyptien* 2nd Series, no. 2 (1881): 55–69.
Abdel-Malek, Anouar. *Egypt, Military Society: The Army Regime, the Left, and Social Change Under Nasser*. Translated by Charles Lam Markmann. Random House, 1968.
Abu Lughod, Janet. *Cairo: 1001 Years of the City Victorious*. Princeton University Press, 1971.
Abul-Magd, Zeinab. *Militarizing the Nation: The Army, Business, and Revolution in Egypt*. Columbia University Press, 2017.
Ackerknecht, Erwin H. "Anticontagionism between 1821 and 1867." *Bulletin of the History of Medicine* 22, no. 5 (1948): 562–93.
'Afifi, Muhammad. *Al-Awqaf wa al-Haya al-Iqtisadiyya fi Misr fi al-'Asr al-'Uthmani*. al-Hay'a al-Misriyya al-'Amma li al-Kitab, 1991.
Ahmad, Mahmud Samir. *Ma'arik al-Miyah al-Muqbila fi al-Sharq al-Awsat*. Dar al-Mustaqbal al-'Arabi, 1991.
Alder, Ken. *Engineering the Revolution: Arms and Enlightenment in France, 1763–1815*. University of Chicago Press, 1997.

Alleaume, Ghislaine. "Hygiène publique et travaux publics: les ingénieurs et l'assainissement du Caire (1882–1907)." *Annales Islamologiques* 20 (1984): 151–82.

Allen, Richard W. "Cairo Drainage Scheme." *Journal of the Royal African Society* 28, no. 112 (1929): 340–46.

'Alluba, 'Ali. *Mabadi' fi al-Siyasa al-Misriyya*. Dar al-Kutub, 1942.

'Amir, 'Abd al-Salam, *Tawa'if al-Hiraf fi Misr, 1805–1914*. Al-Hay'a al-Misriyya al-'Amma li al-Kitab, 1993.

American Water Works Association. *Water Meters—Selection, Installation, Testing, and Maintenance*, Fifth Ed. American Water Works Association, 2012.

Amin, Ahmad. *Qamus al-'Adat wa al-Taqalid wa al-Ta'abir al-Misriyya*. Hindawi, 2013 [1953].

Anand, Nikhil. *Hydraulic City: Water and the Infrastructure of Citizenship in Mumbai*. Duke University Press, 2017.

'Anbar, Muhammad Sadiq. *Dhikra Faqid al-Watan al-Maghfur Lahu Amin al-Rafi'i*. Al-Nahda, 1928.

Arminjon, Pierre. "La Crise Financière Égyptienne Actuelle." *Revue des deux mondes* 77, no. 41 (1907): 198–221.

Arnaud, Jean-Luc. *Le Caire: mise en place d'une ville moderne, 1867–1907: Des intérêts du prince aux sociétés privées*. Actes Sud, 1998.

Artin, Yacoub. "Le Premier Puits Artésien dans la Vallée du Nil." *Bulletin de l'Institut Égyptien* 3rd Series, no. 7 (1896): 29–33.

———. *De l'eau du nil et de l'eau souterraine*. Imprimerie Nationale, 1904.

———. *Essai sur les causes du renchérissement de la vie matérielle au Caire dans le courant du XIXe siècle (1800 à 1907)*. Imprimerie de l'Institut français d'archéologie orientale, 1907.

Badrawi, Malak. *Political Violence in Egypt, 1910–1924: Secret Societies, Plots and Assassinations*. Routledge, 2000.

Bahjat, Nabil. *Badi' Khayri: al-A'mal al-Shi'riyya al-Kamila*. Dar Mirit, 2010.

Bakker, Karen. *Privatizing Water: Governance Failure and the World's Urban Water Crisis*. Cornell University Press, 2010.

Barak, On. *On Time: Technology and Temporality in Modern Egypt*. University of California Press, 2013.

———. *Powering Empire How Coal Made the Middle East and Sparked Global Carbonization*. University of California Press, 2020.

Baran, Paul and Paul Sweezy. *Monopoly Capital: An Essay on the American Economic and Social Order*. Monthly Review Press, 1966.

Barois, Julien. *Rapport sur les recherches du pétrole de la Mer Rouge*. La Caire: Imprimerie Nationale, 1885.

———. "Notice sur le climat du Caire." *Bulletin de l'Institut Égyptien* 2nd Series, no. 10 (1889): 78–210.

———. *Les irrigations en Égypte*. Librairie Polytechnique Ch. Béranger, 1904.

Batty, Michael and Stephen Marshall. "The Evolution of Cities: Geddes, Abercrombie and the New Physicalism." *The Town Planning Review* 80, no. 6 (2009): 551–74.

Bayat, Asef, and Eric Denis. "Who is Afraid of *Ashwaiyyat*? Urban Change and Politics in Egypt." *Environment and Urbanization* 12, no. 2 (2000): 185–99.

Beinin, Joel and Zackary Lockman. *Workers on the Nile: Nationalism, Communism, Islam, and the Egyptian Working Class, 1882–1954*. Princeton University Press, 1988.

Bernet, Claus. "The 'Hobrecht Plan' (1862) and Berlin's Urban Structure." *Urban History* 31, no. 3 (2004): 400–19.

Berque, Jacques. *Egypt: Imperialism and Revolution*. Faber & Faber, 1972.

Bijker, Wiebe E., Thomas Hughes, and Trevor Pinch, eds. *Social Construction of Technological Systems: New Directions in the Sociology and History of Technology*. MIT Press, 1987.

Bitter, H. *Observations hygiéniques sur l'approvisionnement en eau de la nappe souterraine - projété pour la Ville de Tantah par M. I'Ingénieur K. Abel*. Le Caire: Typ. Lithographie A. Costagliola, 1897.

———. "Le choléra et la peste en Égypte." In *Comptes Rendus de Premier Congrès Égyptien de Médecine tenu au Caire du 19 au 23 Décembre 1902*. Imprimerie Nationale, 1904.

Bitter, H. and Henry Blagden. *Filtration Experiments at Alexandria, Egypt: Report on the Efficiency of the "Jewell" Filter*. New York Continental Jewell Filtration Co., 1903.

Bix, Amy. *Girls Coming to Tech!: A History of American Engineering Education for Women*. MIT Press, 2014.

Björkman, Lisa. *Pipe Politics, Contested Waters: Embedded Infrastructures of Millennial Mumbai*. Duke University Press, 2015.

Björkman, Lisa and Andrew Harris. "Engineering Cities: Mediating Materialities, Infrastructural Imaginaries and Shifting Regimes of Urban Expertise." *International Journal of Urban and Regional Research* 42, no. 2 (2018): 244–62.

Black, R. D. Collinson. "W. S. Jevons and the Foundation of Modern Economics." In *The Marginal Revolution in Economics: Interpretation and Evaluation*, edited by R. D. Collinson Black, A. W. Coats, and Craufurd D. W. Goodwin. Duke University Press, 1973.

Bocquet, Denis, Konstantinos Chatzis, and Agnès Sander. "From Free Good to Commodity: Universalizing the Provision of Water in Paris (1830–1930)." *Geoforum* 39, no. 6 (2008): 1821–32.

Boulad, Emile. *Les Tramways du Caire*. Paul Barbey Press, 1919.

Bret, Patrice. "Gunpowder Manufacture in Cairo from Bonaparte to Muhammad Ali: Adaptation, Innovation and the Transfer of Technology, 1798–1820." In *Gunpowder, Explosives and the State: A Technological History*, edited by Brenda Buchanan. Aldershot, Hants, Ashgate Publishing, 2006.

Briggs, Asa. "Samuel Smiles and the Gospel of Work." In *Victorian People: A Reassessment of Persons and Themes, 1851–1867*, edited by Asa Briggs. University of Chicago Press, 1954.

———. "Cholera and Society in the Nineteenth Century." *Past and Present* 19 (1961): 76–96.

Broich, John. *London: Water and the Making of the Modern City*. University of Pittsburgh Press, 2013.

Nathan J. Brown, "The Precarious Life and Slow Death of the Mixed Courts of Egypt," *International Journal of Middle East Studies* 25, no. 1 (1993): 33–52.

Brown, R. H. *History of the Barrage at the Head of the Delta of Egypt*. Cairo: F. Diemer 1896.

Buchanan, Robert. *The Engineers: A History of the Engineering Profession in Britain, 1750–1914*. London: Jessica Kingsley Publishers, 1989.

Callon, Michel. "Society in the Making: The Study of Technology as a Tool for Sociological Analysis." In *The Social Construction of Technological Systems: New Directions in the Sociology and History of Technology*, edited by Wiebe Bijker, Thomas Parke Hughes, and Trevor Pinch. MIT Press, 1987.

Camprubí, Lino. *Engineers and the Making of the Francoist Regime*. MIT Press, 2014.

Caneri, Toussaint. *La Ville du Caire, son Présent et son Avenir: Essai sur la Création d'une Municipalité*. Imprimerie de l'Institut français d'archéologie orientale, 1905.

Castree, Noel. "Neoliberalising nature: The Logics of Deregulation and Reregulation." *Environment and Planning* 40, no. 1 (2008): 131–52.

Chalcraft, John. *The Striking Cabbies of Cairo and Other Stories: Crafts and Guilds in Egypt, 1863–1914*. State University of New York Press, 2005.

Chehata, Tawfik. "La Concession de Service Public." *L'Égypte Contemporaine* 31–32, nos. 195–201 (1941): 205–496.

Cherry, Gordon E. "The Town Planning Movement in the Late Victorian City." *Transactions of the Institute of British Geographers* 4, no. 2 (1979): 306–19.

Claude, Viviane. "Sanitary Engineering as a Path to Town Planning: The Singular Role of the Association Générale des Hygiénistes et Techniciens Municipaux in France and the French-Speaking Countries, 1900–1920." *Planning Perspectives* 4, no. 2 (1989): 153–66.

Clavel, Pierre and Robert Young. ""Civics": Patrick Geddes's Theory of City Development." *Landscape and Urban Planning* 166 (2017): 37–42.

Clerget, Marcel. *Le Caire: étude de géographie urbaine et d'histoire économique*. 2 vols. E. & R. Schindler Press, 1934.

Cole, Juan. *Colonialism and Revolution in the Middle East: Social and Cultural Origins of Egypt's Urabi Movement*. Princeton University Press, 1993.

Coleman, William. "Koch's comma bacillus: the first year." *Bulletin of the History of Medicine* 61, no. 3 (1987): 315–42.

Cooper, Frederick, and Randall M. Packard, eds. *International Development and the Social Sciences: Essays on the History and Politics of Knowledge*. University of California Press, 1997.

Cooper, Timothy. "Peter Lund Simmonds and the Political Ecology of Waste Utilization in Victorian Britain." *Technology and Culture* 52, no. 1 (2011): 21–44.

Corbin, Alain. *The Foul and the Fragrant: Odor and the French Social Imagination*. Translated by Aubier Montaigne. Harvard University Press, 1988.

Corthell, Elmer. "A Recent Trip to Italy, Egypt, Southern France and Spain." *Professional Memoirs, Corps of Engineers, United States Army, and Engineer Department at Large* 7, no. 33 (1915): 253–306.

Cottarel, François. *Bahth Tarshih Miyah al-Nil al-Mu'adda li al-Shurb*. Matba'at Paul Perrier, 1910.

Cronon, William. *Nature's metropolis: Chicago and the Great West*. W. W. Norton and Co., 1991.

Crouch, Melissa L., Heinz E. Jacobs, and Vanessa L. Speight. "Defining Domestic Water Consumption Based on Personal Water Use Activities." *Journal of Water Supply: Research and Technology-AQUA* 70, no. 7 (2021): 1002–11.

Crouchley, A. E. *The Investment of Foreign Capital in Egyptian Companies and Public Debt*. Cairo: Government Press, 1936.

Cushman, Gregory. *Guano and the Opening of the Pacific World: A Global Ecological History*. Cambridge University Press, 2013.

Darwish, Hasan. *Badi' Khayri: al-Azjal al-Badi'a wa al-Alhan al-Rayhaniyya: Dirasa Fanniyya Siyasiyya Ijtima'iyya*. Al-Majlis al-A'la li al-Thaqafa, 2001.

Davis, Mike. *Planet of Slums*. Verso, 2006.

Davison, Graeme. "The City as a Natural System: Theories of Urban Society in Early Nineteenth-Century Britain." In *The Pursuit of Urban History*, edited by Derek Fraser and Anthony Sutcliffe. Edward Arnold, 1983.

Deeb, Marius. *Party Politics in Egypt: The Wafd & its Rivals, 1919–1939*. Ithaca Press for the Middle East Centre, St. Antony's College, Oxford, 1979.

Denis, Eric. "Cairo as Neo-Liberal Capital?: From Walled City to Gated Communities." In *Cairo Cosmopolitan: Politics, Culture, and Urban Space in the New Globalized Middle East*, edited by Diane Singerman and Paul Amar. American University in Cairo Press, 2006.

Dieter, Cheryl A., and Molly A. Maupin. *Public Supply Water Use and Self-Supplied Industrial Water Use in the United States, 2015*. U.S. Geological Survey Open-File Report 2017–1131. Reston, VA: U.S. Geological Survey, 2017.

Derr, Jennifer. *The Lived Nile: Environment, Disease, and Material Colonial Economy in Egypt*. Stanford University Press, 2019.

Dobb, Maurice. *Theories of Value and Distribution since Adam Smith: Ideology and Economic Theory*. Cambridge University Press, 1973.

Dobrowolska, Agnieszka, and Jarosław Dobrowolski. *The Sultan's Fountain: An Imperial Story of Cairo, Istanbul, and Amsterdam*. American University in Cairo Press, 2011.

Drayton, Richard. *Nature's Government: Science, Imperial Britain, and the "Improvement" of the World*. Yale University Press, 2000.

Duarte, Pedro Garcia, and Yann Giraud, eds. *Economics and Engineering: Institutions, Practices, and Cultures*. Duke University Press, 2020.

Duffield, Roberta. "Military Capitalism: The Economic and Security Logics of Egypt's New Administrative Capital." In *Cairo Securitized: Reconceiving Urban Justice and Social Resilience*, edited by Paul Amar. American University in Cairo Press, 2024.

Dupuit, Jules. *De L'Utilité et de sa Mesure*. Edited by Mario de Bernardi. La Riforma Sociale, 1933.

———. "On Tolls and Transportation Charges." *International Economic Papers* 11 (1962): 7–31.

———. "On the Measurement of the Utility of Public Works." In *Readings in Welfare Economics*, edited by Kenneth J. Arrow and Tibor Scitovsky, translated by R. H. Barback. Allen and Unwin, 1969.

Eid, Alfred. *La Fortune Immobilière de l'Égypte et sa Dette Hypothécaire: Propriété Rurale, Propriété Urbaine*. F. Alcan, 1907.

Ekelund, Robert B., and Robert F. Hébert. *Secret Origins of Modern Microeconomics: Dupuit and the Engineers*. University of Chicago Press, 1999.

———. "Retrospectives: The Origins of Neoclassical Microeconomics." *The Journal of Economic Perspectives* 16, no. 3 (2002): 197–215.

Ellet, Charles. *An Essay on the Laws of Trade: In Reference to the Works of Internal Improvement in the United States*. Richmond: P. D. Bernard, 1839.

———. *The Laws of Trade Applied to the Determination of the Most Advantageous Fare for Passengers on Rail Roads*. Philadelphia, 1840.

Elshahed, Mohamed. *Cairo Since 1900: An Architectural Guide*. American University in Cairo Press, 2020.

Elshakry, Marwa. *Reading Darwin in Arabic, 1860–1950*. Chicago University Press, 2013.

Etner, François. *Histoire du calcul économique en France*. Éditions Economica, 1987.

Evans, Richard J. *Death in Hamburg: Society and Politics in the Cholera Years, 1830–1910*. Clarendon Press, 1987.

Fahmy, Khaled. "An Olfactory Tale of Two Cities." In *Historians in Cairo: Essays in Honor of George Scanlon*, edited by Jill Edwards. The Americas University in Cairo Press, 2002.

———. "Modernizing Cairo: A Revisionist Narrative." In *Making Cairo Medieval*, edited by Nezar Alsayyad, Irene Bierman, and Nasser Rabbat. Lexington Press, 2005.

———. *In Quest of Justice: Islamic Law and Forensic Medicine in Modern Egypt*. University of California Press, 2018.

Fahmi, Zaki. *Safwat al-'Asr fi Tarikh wa Rusum Mashahir Rijal Misr*. Hindawi, 2013 [1926].

Fadli, Fodil and Magda Sibley. "The Historic *Hammāms* of Cairo." *Journal of Architectural Conservation* 14, no. 3 (2008): 59–80.

Farmer, Tessa. "Willing to Pay: Competing Paradigms about Resistance to Paying for Water Services in Cairo, Egypt." *Middle East Law and Governance* 9 (2017): 3–19.

Ferguson, Eugene. *Engineering and the Mind's Eye*. MIT Press, 1992.

Forty, Adrian. *Objects of Desire: Design and Society since 1750*. Thames and Hudson, 1986.

Foster, John Bellamy. "Marx's Theory of Metabolic Rift: Classical Foundations for Environmental Sociology." *American Journal of Sociology* 105, no. 2 (1999): 366–405.

Foucart, George. "Le Caire et la speculation." *La nouvelle revue* 47 (1907): 487–504.

Frapier, Christel, and Simon Vaillant. "The Organization of the Hennebique Firm in the Countries of the Mediterranean Basin: Establishment and Communications Strategy." In *Building Beyond the Mediterranean: Studying the Archives of European Businesses (1860–1970)*, edited by Claudine Piaton, Ezio Godoli, and David Peyceré. Honoré Clair, 2012.

Galbraith, John Kenneth. *A Short History of Financial Euphoria*. Penguin Books, 1990.

Gandy, Matthew. "The Paris Sewers and the Rationalization of Urban Space." *Transactions of the Institute of British Geographers* 24, no. 1 (1999): 23–44.

———. *The Fabric of Space: Water, Modernity, and the Urban Imagination*. MIT Press, 2014.

Gesink, Indira Falk. *Islamic Reform and Conservatism: Al-Azhar and the Evolution of Modern Sunni Islam*. I. B. Tauris, 2010.

Gilberg, Mark. "Alfred Lucas: Egypt's Sherlock Holmes." *Journal of the American Institute for Conservation* 36, no. 1 (1997): 31–48.

Gold, Meira. "Ancient Egypt and the Geological Antiquity of Man, 1847–1863." *History of Science* 57, no. 2 (2019): 194–230.

Goldschmidt, Arthur. *Biographical Dictionary of Modern Egypt*. Lynne Rienner Publishers, 2000.

Goubert, Jean-Pierre. *The Conquest of Water: The Advent of Health in the Industrial Age*. Princeton University Press, 1989.

Graham, Steve, and Simon Marvin. *Splintering Urbanism: Networked Infrastructures, Technological Mobilities and the Urban Condition*. Routledge, 2001.

Graham-Leigh, John. *London's Water Wars: The Competition for London's Water Supply in the Nineteenth Century*. Francis Boutle Publishers, 2000.

Grant, James. "Cholera." *Virginia Medical Monthly* 12, no. 12 (1886): 724–34.

Guerville, Amédée Baillot de. *La nouvelle Égypte*. Librairie Universelle, 1905.

Günel, Gökçe. *Spaceship in the Desert: Energy, Climate Change, and Urban Design in Abu Dhabi*. Duke University Press, 2019.

Hamlin, Christopher. *Public Health in the Age of Chadwick*. Cambridge University Press, 1998.

———. "Edwin Chadwick and the Engineers, 1842–1854: Systems and Antisystems in the Pipe-and-Brick Sewers Wars." In *Water-Supply and Public Health Engineering*, edited by Denis Smith. Ashgate, 1999.

Hanley, Susan. "Urban Sanitation in Preindustrial Japan." *The Journal of Interdisciplinary History* 18, no. 1 (1987): 1–26.

Hanna, Nelly. *An Urban History of Būlāq in the Mamluk and Ottoman Periods*. Institut Français d'Archéologie Orientale, 1983.

Harvey, David. *Paris, Capital of Modernity*. Routledge, 2003.

———. *Rebel Cities: From the Right to the City to the Urban Revolution*. Verso, 2012.

Haswell, C. "Town Planning and Housing in Cairo." *Garden Cities & Town Planning* 11 (1921): 256–58.

Hastaoglou-Martinides, Vilma. "The Building of Istanbul Docks 1870–1910. New Entrepreneurial and Cartographic Data." *A/Z ITU Journal of Faculty of Architecture* 8, no. 1 (2011): 85–99.

Havinden, Michael, and David Meredith. *Colonialism and Development: Britain and Its Tropical Colonies 1850–1960*. Routledge, 1996.

Heilbroner, Robert, and William Milberg. *The Crisis of Vision in Modern Economic Thought*. Cambridge University Press, 1995.

Henderson, W. O. "Walther Rathenau: A Pioneer of the Planned Economy." *The Economic History Review* 4, no. 1 (1951): 98–108.

Hobrecht, James. *Die Canalisation von Berlin*. Berlin: Ernst & Korn, 1884.

Hobson, A. J. *Imperialism: A Study*. James Pott & Company, 1902.

Hodge, A. Trevor. *Roman Aqueducts and Water Supply*. Duckworth, 1992.

Hodge, Joseph Morgan. *Triumph of the Expert: Agrarian Doctrines of Development and the Legacies of British Colonialism*. Ohio University Press, 2007.

Hollings, Albright, ed. *The Life of Sir Collin C. Scott-Moncrieff*. John Murray, 1917.

Home, Robert. "British Colonial Town Planning in the Middle East: The Work of W. H. McLean." *Planning History* 12, no. 1 (1990): 4–9.

———. *Of Planting and Planning: The Making of British Colonial Cities*. E & F. N. Spon, 1997.

Horner, Leonard. "An Account of Some Recent Researches Near Cairo, Undertaken with the View of Throwing Light upon the Geological History of the Alluvial Land of Egypt." *Philosophical Transactions of the Royal Society of London* 145 (1855): 105–38.

———. "An Account of Some Recent Researches Near Cairo, Undertaken with the View of Throwing Light upon the Geological History of the Alluvial Land of Egypt." *Philosophical Transactions of the Royal Society of London* 148 (1858): 53–92.

Howard-Jones, Norman. *The Scientific Background of the International Sanitary Conferences, 1851–1938*. World Health Organization, 1975.

Hubert, Bonin. *History of the Suez Canal Company, 1858–1960: Between Controversy and Utility*. Librairie Droz, 2010.

Hugo, Victor. *Les Misérables*. Translated by Christine Donougher. Penguin, 2013.

Hunt, E. K., and Mark Lautzenheiser. *History of Economic Thought: A Critical Perspective*. Routledge, 2011.

Hunter, Robert. *Egypt Under the Khedives, 1805–1879: From Household Government to Modern Bureaucracy*. The University of Pittsburgh Press, 1984.

Husayni, Mahmud al-. *Al-Asbila al-'Uthmaniyya bi Madinat al-Qahira, 1517–1798*. Maktabat Madbuli, 1988.

Ilbert, Robert. *Héliopolis, le Caire, 1905–1922: genèse d'une ville*. Centre Nationale de la Recherche Scientifique, 1981.

Ismail, Abdel-Mawla. "Drinking Water Protests in Egypt and the Role of Civil Society." *Transnational Institute* (2009). https://www.tni.org/files/drinking_water_protest_in_egypt_by_abdel_mawla.pdf.

Ismail, Shehab. "Epicures and Experts: The Drinking Water Controversy in British Colonial Cairo." *Arab Studies Journal* 26, no. 2 (2018): 8–43.

Issawi, Charles, and Felix Rosenfeld. "Company Profits in Egypt, 1929-1939." *L'Égypte Contemporaine* 32, no. 204 (1941): 679–86.

Jabarti, 'Abd al-Rahman al-. *'Aja'ib al-Athar fi al-Tarajim wa al-Akhbar*. 4 vols. Matba'at Dar al-Kutub, 1998.

Jacob, Wilson. *Working Out Egypt: Effendi Masculinity and Subject Formation in Colonial Modernity, 1870–1940*. Duke University Press, 2011.

Jakes, Aaron. *Egypt's Occupation: Colonial Economism and the Crisis of Capitalism*. Stanford University Press, 2020.

James, Charles Carkeet. *Drainage Problems of the East*. *The Times of India* Office, 1906.

———. "The Main Drainage of Cairo." *Minutes of Proceeding—Institution of Civil Engineers* 202 (1916): 57–95.

Jevons, William Stanley. *The Theory of Political Economy*. Palgrave Macmillan, 2013 [1871].

———. *Letters and Journals of W. Stanley Jevons*. Edited by Harriet A. Jevons. London: Macmillan and Co., 1886.

Judd, J. W. "Report On a Series of Specimens of the Deposits of the Nile Delta, Obtained by the Recent Boring Operations." *Proceedings of the Royal Society of London* 39 (1885): 213–27.

———. "Second Report on a Series of Specimens of the Deposits of the Nile Delta, Obtained by Boring Operations Undertaken by the Royal Society." *Proceedings of the Royal Society of London* 61 (1897): 32–40.

Kaufmann, Paul. "Sur le Prétendu Nil Vert." *Revue d'Égypte* 4 (1897): 105–13.

Kellet, J. R. "Municipal Socialism, Enterprise and Trading in the Victorian City." *Urban History Yearbook* 1, no. 5 (1978): 36–45.

Khalil, Omnia. "Cairo's Incarcerated Geographies." *Middle East Report*, no. 312 (2024).

Khalili, Laleh. *Sinews of War and Trade: Shipping and Capitalism in the Arabian Peninsula*. Verso, 2020.

Khuri-Makdisi, Ilham. *The Eastern Mediterranean and the Making of Global Radicalism, 1860–1914*. University of California Press, 2010.

Kilani, Muhammad Sayyid. *Tiram al-Qahira*. Cairo: Al-Hai'a al-'Amma li Qusur al-Thaqafa, 2010 [1968].

Kindleberger, Charles and Robert Aliber. *Manias, Panics, and Crashes: A History of Financial Crises*. 5th ed. John Wiley & Sons, Inc., 2005 [1978].

Knoedler, Janet and Anne Mayhew. "Thorstein Veblen and the Engineers: A Reinterpretation." *History of Political Economy* 31, no. 2 (1999): 255–72.

Koen, Billy Vaughn. *Discussion of the Method: Conducting the Engineer's Approach to Problem Solving*. Oxford University Press, 2003.

Krämer, Gudrun. *The Jews in Modern Egypt, 1914–1952*. I. B. Tauris, 1989.

Kranakis, Eda. "Social Determinants of Engineering Practice: A Comparative View of France and America in the Nineteenth Century." *Social Studies of Science* 19, no. 1 (1989): 5–70.

———. "Fixing the Blame: Organizational Culture and the Quebec Bridge Collapse." *Technology and Culture* 45, no. 3 (2004): 487–518.

Kuhnke, LaVerne. *Lives at Risk: Public Health in Nineteenth-Century Egypt*. Berkeley: University of California Press, 1990.

Kupferschmidt, Uri M. "Who Needed Department Stores in Egypt? From Orosdi-Back to Omar Effendi." *Middle Eastern Studies* 43, no. 2 (2007): 175–92.

Kurz, Heinz D. *Economic Thought: A Brief History*. Columbia University Press, 2016.

Labib, Malak. *Recenser l'Égypte: Dette publique et politiques de quantification à l'ère impériale (1875–1922)*. Institut Français d'Archéologie Orientale, 2024.

Landes, David. *Bankers and Pashas: International Finance and Economic Imperialism in Egypt*. Harvard University Press, 1958.

Lane-Poole, Stanley. *The Story of Cairo*. London: J. M. Dent & Co., 1906.

Laporte, Dominique. *History of Shit*. Translated by Nadia Benabid and Rodolphe el-Khoury. MIT Press, 2000 [1978].

Lardner, Dionysius. *Railway Economy*. London: Taylor, Walton, & Maberly, 1850.

Latham, Baldwin. *Sanitary Engineering: A Guide to the Construction of Works of Sewerage and House Drainage*. 2nd ed. London: E. & F. N. Spon, 46, Charing Cross, 1878.

Latour, Bruno. *Aramis, or The Love of Technology*. Harvard University Press, 1996.

Law, John and Michel Callon. "Engineering and Sociology in a Military Aircraft Project: A Network Analysis of Technological Change." *Social Problems* 35, no. 3 (1988): 284–97.

Layton, Edwin. *The Revolt of the Engineers: Social Responsibility and the American Engineering Profession*. Johns Hopkins University Press, 1986 [1971].

Lenin, Vladimir. *Imperialism: The Highest Stage of Capitalism*. Penguin Books, 2010 [1916].

Le Strat, Anne. "Discussion – the Remunicipalization of Paris's Water Supply Service: A Successful Reform." *Water Policy* 16, no. 1 (2014): 197–204.

Liebig, Justus von. *Letters on Modern Agriculture*. New York: John Wiley, 1859.

———. *Two Letters on the Subject of the Utilization of the Metropolitan Sewage: Addressed to the Lord Mayor of London*. London: W. H. Collingridge, 1865.

Light, Jennifer. "When Computers Were Women." *Technology and Culture* 40, no. 3 (1999): 455–83.

Low, Michael Christopher. *Imperial Mecca: Ottoman Arabia and the Indian Ocean Hajj*. Columbia University Press, 2020.

Luxemburg, Rosa. *The Accumulation of Capital*. Translated by Agnes Schwarzschild. Routledge, 2003 [1913].

Madani, Al-Sayyid Muhammad. *Mas'uliyyat al-Dawla 'an A'amaliha al-Musharra'a: (Al-Qawanin wa al-Lawa'h) fi al-Qanun al-Misri—Dirasa Muqarana*. Al-Mabta'a al-'Alamiyya, 1952.

Madhavan, Guru. *Applied Minds: How Engineers Think*. W. W. Norton, 2016.

Mahboub, M. Sabry. "Cairo: Some Notes on its History, Characteristics and Town Plan." *Town Planning Institute* 21 (1934/35), 288–302.

Malm, Andreas. *Fossil Capital: The Rise of Steam Power and the Roots of Global Warming*. Verso, 2016.

Magnusson, Roberta J. *Water Technology in the Middle Ages: Cities, Monasteries, and Waterworks after the Roman Empire*. Johns Hopkins University Press, 2001.

Maqrizi, Taqi al-Din Ahmad b. 'Ali b. 'Abd al-Qadir al-. *Al-Maw'aiz wa al-I'tibar fi Dhikr al-Khitat wa al-Athar*. 3 vols. Maktabat Madbuli, 1998.

Marx, Karl. *Capital: A Critique of Political Economy*. 3 vols. Penguin in association with New Left Review, 1990 [1867].

Mårald, Erland. "Everything Circulates: Agricultural Chemistry and Recycling Theories in the Second Half of the Nineteenth Century." *Environment and History* 8, no. 1 (2002): 65–84.

Maunier, René. "La Vie Municipale en Egypte." In *Premier Congrès International et Exposition Comparée des Villes*. Part II, 33–39. Union Internationale des Villes, 1913.

Mays, Larry W., ed. *Ancient Water Technologies*. Springer, 2010.

McDonald, David. "Remunicipalization: The Future of Water Services?" *Geoforum* 91 (2018): 47–56.

McLean, William. "Local Government and Town Development in Egypt." *The Town Planning Review* 7, no. 2 (1917): 83–97.

Meijer, Roel. *The Quest for Modernity: Secular Liberal and Left-Wing Political Thought in Egypt, 1945–1958*. Routledge Curzon, 2002.

Meller, Helen. *Patrick Geddes: Social Evolutionist and City Planner*. Routledge, 1990.

Melosi, Martin. "The Place of the City in Environmental History." *Environmental History Review* 17, no. 1 (1993): 1–23.

———. *The Sanitary City: Urban Infrastructures in America from the Colonial Times to the Present*. Johns Hopkins University Press, 1999.

Mieton, Fredrik. *Electrical Palestine: Capital and Technology from Empire to Nation*. University of California Press, 2018.

Mill, John Stuart. *Essays on Economics and Society.* University of Toronto Press, 1967.

Millington, Nate. "Producing Water Scarcity in São Paulo, Brazil: The 2014–2015 Water Crisis and the Binding Politics of Infrastructure." *Political Geography* 65 (July, 2018): 26–34.

Millington, Nate and Suraya Scheba. "Day Zero and The Infrastructures of Climate Change: Water Governance, Inequality, and Infrastructural Politics in Cape Town's Water Crisis." *International Journal of Urban and Regional Research* 45, no. 1 (2021): 116–32.

Millward, Robert. "Emergence of Gas and Water Monopolies in Nineteenth-century Britain: Contested Markets and Public Control." In *New Perspectives on the Late Victorian Economy*, edited by James Foreman-Peck. Cambridge University Press, 1991.

———. "The Political Economy of Urban Utilities." In *The Cambridge Urban History of Britain*, edited by Martin Daunton. Cambridge University Press, 2001.

Milner, Alfred. *England in Egypt.* London: Edward Arnold, 1894 [1892].

Milonakis, Dimitris, and Ben Fine. *From Political Economy to Economics: Method, the Social and the Historical in the Evolution of Economic Theory.* Routledge, 2009.

Mitcham, Carl. *Thinking Through Technology: The Path Between Engineering and Philosophy.* University of Chicago Press, 2014.

Mitchell, Timothy. *Rule of Experts: Egypt, Techno-Politics, Modernity.* University of California Press, 2002.

———. *Carbon Democracy: Political Power in the Age of Oil.* Verso, 2011.

Moore, Clement Henry. *Images of Development: Egyptian Engineers in Search of Industry.* MIT Press, 1980.

Morgan, Emily Kathryn. ""True Types of the London Poor:" Adolphe Smith and John Thompson's Street Life in London." PhD diss., University of Arizona, 2012.

Mosca, Manuela. "On the Origins of the Concept of Natural Monopoly: Economies of Scale and Competition." *European Journal of the History of Economic Thought* 15, no. 2 (2008): 317–53.

Mostafa, Saleh Lamei. "The Cairene Sabil: Form and Meaning." *Muqarnas* 6 (1989): 33–42.

Mubarak, ʿAli. *Al-Khitat al-Tawfiqiyya al-Jadida li Misr al-Qahira.* 20 vols. Cairo: Matbaʿat Bulaq, 1888.

Mukerji, Chandra. *Impossible Engineering: Technology and Territoriality on the Canal Du Midi.* Princeton University Press, 2009.

Musallami, Ibrahim ʿAbdalla al-. *Ahmad Hilmy: sajin al-hurriyya wa al-sahafa.* Cairo: Al-Hay'a al-Misriyya al-ʿAmma li al-Kitab, 1993.

Nazmi, ʿAbd al-ʿAziz, ed. *Bahth Wajiz ʿan Sharikat al-Ihtikar bi Misr.* Matbaʿat al-Fustat, 1934.

Nile Projects Commission. *Minutes of the Discussion Between the Nile Projects Commission and the Committee of Egyptian Engineers, at the Public Meeting Held on 22nd June 1920.* Cairo: 1920.

Nitzan, Jonathan. "Differential Accumulation: Towards a New Political Economy of Capital." *Review of International Political Economy* 5, no. 2 (1998): 169–216.

Noble, David. *America by Design: Science, Technology and the Rise of Corporate Capitalism*. Alfred A. Knopf, 1977.

Noyes, Alexander. "A Year After the Panic of 1907." *The Quarterly Journal of Economics* 23, no. 2 (1909): 185–212.

Nubarian, Nubar. *Mudhakirat Nubar Pasha*. Dar al-Shuruq, 2009.

Odlyzko, Andrew. "Collective Hallucinations and Inefficient Markets: The British Railway Mania of the 1840s." University of Minnesota manuscript (January 15, 2010). https://ssrn.com/abstract=1537338 or http://dx.doi.org/10.2139/ssrn.1537338.

———. "Dionysius Lardner, the Denigrated Sage of Early Railways." University of Minnesota manuscript (August 30, 2019). https://ssrn.com/abstract=3445470 or http://dx.doi.org/10.2139/ssrn.3445470.

OECD/Lincoln Institute of Land Policy, PKU-Lincoln Institute Center. *Global Compendium of Land Value Capture Policies*. OECD Regional Development Studies. OECD Publishing, 2022. https://doi.org/10.1787/4f9559ee-en.

Ogawa, Mariko. "Uneasy Bedfellows: Science and Politics in the Refutation of Koch's Bacterial Theory of Cholera." *Bulletin of the History of Medicine* 74, no. 4 (2000): 671–707.

Oldenziel, Ruth. *Making Technology Masculine: Men, Women, and Modern Machines in America, 1870–1945*. Amsterdam University Press, 1999.

Owen, Roger. *Cotton and the Egyptian Economy, 1820–1914: A Study in Trade and Development*. The Clarendon Press, 1969.

———. "The Cairo Building Industry and the Building Boom of 1897 to 1907." In *Colloque internationale sur l'histoire du Caire*. Ministry of Culture of the Arab Republic of Egypt, 1972.

———. *The Middle East in the World Economy*. I. B. Tauris, 2009 [1981].

Parodi, H. D. "La fabrication des engrais au Caire." *L'Égypte Contemporaine* 1 (1910): 424–31.

Pasquier, Émilie. *La Société des Eaux du Caire: Une Histoire de l'Équipement en Infrastructures Hydrauliques de la Capitale Égyptienne, 1865–1954*. Éditions Ithaka, 2023.

Pauty, Edmond. *Les Hammams du Caire*. Imprimerie de l'Institut Français d'Archéologie orientale, 1933.

Perelman, Michael. *Railroading Economics: The Creation of the Free Market Mythology*. Monthly Review Press, 2006.

Peterson, Jon. "The Impact of Sanitary Reform upon American Urban Planning, 1840–1890." *Journal of Social History* 13, no. 1 (1979): 83–103.

Phalen, William J. *How the Telegraph Changed the World*. McFarland & Company, 2014.

Picon, Antoine. *L'Invention de l'ingénieur moderne: l'École des Ponts et Chaussées, 1747–1851*. Presses de l'École nationale des Ponts et chaussées, 1992.

———. "Engineers and Engineering History: Problems and Perspectives." *History and Technology* 20, no. 4 (2004): 421–36.

———. "The Engineer as Judge: Engineering Analysis and Political Economy in Eighteenth Century France." *Engineering Studies* 1, no. 1 (2009): 19–34.

Pike, David L. *Subterranean Cities: The World Beneath Paris and London, 1800–1945*. Cornell University Press, 2005.

Piper, Karen. *The Price of Thirst: Global Water Inequality and the Coming Chaos*. University of Minnesota Press, 2014.

Piquet, Caroline. "The Suez Company's Concession in Egypt, 1854–1956: Modern Infrastructure and Local Economic Development." *Enterprise & Society* 5, no. 1 (2004): 107–27.

Poffandi, Stefano. *Indicateur égyptien administratif et commercial*. Imprimerie Générale A. Maurés, 1902.

Porter, Theodore M. "Objectivity and Authority: How French Engineers Reduced Public Utility to Numbers." *Poetics Today* 12, no. 2 (1991): 245–65.

———. *Trust in Numbers: The Pursuit of Objectivity in Science and Public Life*. Princeton University Press, 1995.

Powell, Eve Troutt. *A Different Shade of Colonialism: Egypt, Great Britain, and the Mastery of the Sudan*. University of California Press, 2003.

Pyramid. "The Drainage of Cairo." *The Provincial Medical Journal* 14, no. 126 (1895): 298–301.

Quinn, William and John Turner. *Boom and Bust: A Global History of Financial Bubbles*. Cambridge University Press, 2020.

Rafi'i, 'Abd al-Rahman al-. *'Asr Imsa'il*. Dar al-Ma'arif, 1987 [1932].

———. *Mustafa Kamil: Ba'ith al-Haraka al-Wataniyya*. Dar al-Ma'arif, 1999 [1939].

———. *Muhammad Farid: Ramz al-Ikhlas wa al-Tadhya*. Dar al-Ma'arif, 1984 [1941].

———. *Thawrat 1919*. Dar al-Ma'arif, 1987 [1946].

Rafaat, Samir. *Maadi (1904–1962): Society and History in a Cairo Suburb*. The Palm Press, 1994.

Raymond, André. *Cairo*. Translated by Willard Wood. Harvard University Press, 2000.

Reid, Donald. *Paris Sewers and Sewermen: Realities and Representations*. Harvard University Press, 1991.

Reid, Donald Malcolm. *Whose Pharaohs? Archeology, Museums, and Egyptian National Identity from Napoleon to World War I*. University of California Press, 2002.

Reimer, Michael. "Contradiction and Consciousness in Ali Mubarak's Description of al-Azhar." *International Journal of Middle East Studies* 29, no. 1 (1997): 53–69.

———. "Urban Government and Administration in Egypt, 1805–1914." *Die Welt des Islams* 39, no. 3 (1999): 289–318.

Reyburn, Wallace. *Flushed with Pride: The Story of Thomas Crapper*. Polperro Heritage Press, 2009.

Reynolds, Nancy. *A City Consumed: Urban Commerce, the Cairo Fire, and the Politics of Decolonization in Egypt*. Stanford University Press, 2012.

Rizq, Yunan Labib. *Tarikh al-Wizarat al-Misriyya, 1878–1953*. Al-Ahram, 1975.

Rogaski, Ruth. *Hygienic Modernity: Meanings of Health and Disease in Treaty-Port China*. University of California Press, 2014.

Rogers, G. F. C. *The Nature of Engineering: A Philosophy of Technology*. The Macmillan Press, 1983.

Ryzova, Lucie. *The Age of the Efendiyya: Passages to Modernity in National-Colonial Egypt*. Oxford University Press, 2014.

Saadeh Bey, Dr. "Motor Progress in the Land of the Pyramids." *The Horseless Age*, August 21, 1907.

Sabett, Mahgoub [*sic*]. "L'hygiène publique en Égypte." In *Œuvres du congrès national égyptien tenu à Bruxelles le 22, 23, 24 Septembre 1910*. St. Catherine Press, 1911.

Sa'd, [Ahmad] Sadiq. *Ma'sat al-Tamwin*. Dar al-Qarn al-'Ishrin, 1945.

———. *Mushkilat al-Fallah*. Dar al-Qarn al-'Ishrin, 1945.

Sa'id, Naffusa Zakariyya. *Tarikh al-Da'wa ila al-'Amiyya wa Athariha fi Misr*. Dar Nashr al-Thaqafa, 1964.

Said, Salah Zaky. *Nineteenth Century Houses of Old Cairo Quarters*. al-Majlis al-A'la li al-Thaqafa, 2009.

Saito, Kohei. *Karl Marx's Ecosocialism: Capitalism, Nature, and the Unfinished Critique of Political Economy*. Monthly Review Press, 2017.

Sandwith, Fleming Mant. *Egypt as a Winter Resort*. London: Kegan Paul, Trench & Co., 1889.

Saul, Samir. *La France et l'Égypte de 1882 à 1914: Intérêts économiques et implications politiques*. Institut de la gestion publique et du développement économique, 1997.

Schaffer, Simon. "Babbage's Intelligence: Calculating Engines and the Factory System." *Critical Inquiry* 21, no. 1 (1994): 203–27.

Schölch, Alexander. *Egypt for the Egyptians! The Socio-Political Crisis in Egypt 1878–82*. Ithaca Press, 1981.

Schultz, Stanley K., and Clay McShane. "To Engineer the Metropolis: Sewers, Sanitation, and City Planning in Late-Nineteenth-Century America." *Journal of American History* 65, no. 2 (1978): 389–411.

Schwartz, Pedro. "John Stuart Mill and Laissez Faire: London Water." *Economica* 33 (1966): 71–83.

Scott, Adam. "The Liernur Sewerage System at Dordrecht (Holland)." *Journal of the Society of Arts* 24 (1876): 667–70.

Scott, James Harry. *The Law Affecting Foreigners in Egypt: As the Result of the Capitulations, with an Account of their Origin and Development.* William Green and Sons, 1907.

Seikaly, Samir. "Prime Minister and Assassin: Buṭrus Ghālī and Wardānī." *Middle Eastern Studies* 13, no. 1 (1977): 112–23.

Shadwell, Arthur. *The London Water Supply.* New York and Bombay: Longmans, Green, and Co., 1899.

Shakry, Omnia El. "Cairo as Capital of Socialist Revolution?" In *Cairo Cosmopolitan: Politics, Culture, and Urban Space in the New Globalized Middle East.* Edited by Diane Singerman and Paul Amar, 73–98. American University in Cairo, 2009.

Shalabi, Ahmad. *Al-Hukm al-Mahalli wa al-Majalis al-Baladiyya fi Misr.* ʻAlam al-Kutub, 1987.

Shapin, Steven, and Simon Schaffer. *Leviathan and the Air-pump: Hobbes, Boyle, and the Experimental Life.* Princeton University Press, 1989.

Shaʻrawi, Huda. *Mudhakirat Huda Shaʻrawi.* Hindawi, 2017.

Sharubim, Mikha'il. *Al-Kafi fi Tarikh Misr al-Qadim wa al-Hadith.* 4 vols. Dar al-Kutub, 1998.

Shawkat, Yahia. *Egypt's Housing Crisis.* American University in Cairo Press, 2020.

Shokr, Ahmad. *Harvests of Liberation: Cotton, Capitalism, and the End of Empire in Egypt.* Stanford University Press, 2025.

Shone, Isaac. *The Main Drainage of the Houses of Parliament, Westminster.* London: E. & F. N. Spon, 1887.

Sibaʻi, Bashir al-. "Sadiq Saʻd (Abridged Biography)." *Adab Wa Nawd* 6, no. 44 (1989): 29–32.

Simmons, Dana. "Waste Not, Want Not: Excrement and Economy in Nineteenth-Century France." *Representations* 96, no. 1 (2006): 73–98.

Sims, David. *Understanding Cairo: The Logic of a City Out of Control.* The American University in Cairo Press, 2010.

———. *Egypt's Desert Dreams: Development or Disaster?* The American University in Cairo Press, 2015.

Sladen, Douglas. *Oriental Cairo: The City of the "Arabian Nights."* J. B. Lippincott Company, 1911.

Smil, Vaclav. *Enriching the Earth: Fritz Haber, Carl Bosch, and the Transformation of World Food Production.* The MIT Press, 2001.

Smith, Bowden. "The Cairo Sewage Farm at Gebel el Asfar." *The Agricultural Journal of Egypt* 3, no. 1 (1913): 23–27.

———. "The Subsoil of Cairo." *Cairo Scientific Journal* 8, nos. 97 and 98 (1914): 239–50.

Smith, Cecil. "The Longest Run: Public Engineers and Planning in France." *The American Historical Review* 95, no. 3 (1990): 657–92.

Sohrabi, Nader. *Revolution and Constitutionalism in the Ottoman Empire and Iran.* Cambridge University Press, 2011.

Stabile, Donald R. "Veblen and the Political Economy of the Engineer: The Radical Thinker and Engineering Leaders Came to Technocratic Ideas at the Same Time." *The American Journal of Economics and Sociology* 45, no. 1 (1986): 41–52.

Stamatopoulou-Robbins, Sophia. *Waste Siege: The Life of Infrastructure in Palestine.* Stanford University Press, 2019.

Stigler, George. "The Economists and the Problem of Monopoly." *The American Economic Review* 72, no. 2 (1982): 1–11.

Stone Jr., Brian. *Radical Adaptation: Transforming Cities for a Climate Changed World.* Cambridge University Press, 2024.

Stromquist, Shelton. "Municipal Socialism." In vol 2. of *The Cambridge History of Socialism,* edited by Marcel van der Linden. Cambridge University Press, 2022.

Sutcliffe, Anthony. *Towards the Planned City: Germany, Britain, the United States, and France, 1780–1914.* Basil Blackwell, 1981.

Swyngedouw, Erik, Maria Kaïka, and Esteban Castro. "Urban Water: A Political-Ecology Perspective." *Built Environment* 28, no. 2 (2002): 124–37.

Tarr, Joel. "The Evolution of the Urban Infrastructure in the Nineteenth and Twentieth Centuries." In *Perspectives on Urban Infrastructure,* edited by Royce Hanson. National Academy Press, 1984.

———. "The Separate vs. Combined Sewer Problem: A Case Study in Urban Technology and Design Choice." In *Water-Supply and Public Health Engineering,* edited by Denis Smith. Ashgate, 1999.

Tayia, Ahmed, Alexandra M. Collins, and Michael Gilmont. "The Role of Virtual-Water Decoupling in Achieving Food–Water Security: Lessons from Egypt, 1962–2013." *Water International* 47, no. 7 (2022): 1118–39.

Taylor, Philip, ed. *Egyptian Company Manual.* The African World Limited, 1911.

Taylor, Vanessa, and Frank Trentmann. "Liquid Politics: Water and the Politics of Everyday Life in the Modern City." *Past and Present* 211, no. 1 (2011): 199–241.

Tignor, Robert. *Modernization and British Colonial Rule in Egypt.* Princeton University Press, 1966.

———. "Foreign Capital, Foreign Communities, and the Egyptian Revolution of 1952." In *Egypt from Monarchy to Republic: Reassessment of Revolution and Change,* edited by Shimon Shamir. Routledge, 1995.

Todd, John. *Political Economy: A Handbook of Economics and Public Finance for Egyptian Students.* William Hodge and Co., 1910.

Tukhi, Nabil al-. *Tawa'if al-Hiraf fi Madinat al-Qahira.* Al-Hay'a al-Misriyya al-'Amma li al-Kitab, 2009.

Trentmann, Frank, and Vanessa Taylor. "From Users to Consumers: Water Politics in Nineteenth-Century London." In *The Making of the Consumer: Knowledge, Power and Identity in the Modern World,* edited by Frank Trentmann. Oxford University Press, 2006.

Trentmann, Frank. *Empire of Things: How We Became a World of Consumers, from the Fifteenth Century to the Twenty-first.* Allen Lane, 2016.

Tunçer, Ali. *Sovereign Debt and International Financial Control: The Middle East and the Balkans, 1870–1914*. Palgrave Macmillan, 2015.

Tynan, Nicola. "Mill and Senior on London's Water Supply: Agency, Increasing Returns, and Natural Monopoly." *Journal of the History of Economic Thought* 29, no. 1 (2007): 49–65.

Tyrwhitt, Jaqueline, ed. *Patrick Geddes in India*. Lund Humphries, 1947.

Valdovinos, Joyce. "The Remunicipalization of Parisian Water Services: New Challenges for Local Authorities and Policy Implications." *Water International* 37, no. 2 (2012): 107–20.

Vallet, Jean. *Contribution à l'étude de la condition des ouvriers de la grande industrie au Caire*. Imprimerie Valentinoise, 1911.

Veblen, Thorstein. *The Theory of Business Enterprise*. Charles Scribner's Sons, 1915 [1904].

———. *The Engineers and the Price System*. B. W. Huebsch, Inc., 1921.

Vincenti, Walter. *What Engineers Know and How they Know It: Analytical Studies from Aeronautical History*. Johns Hopkins University Press, 1990.

Vitalis, Robert. *When Capitalists Collide: Business Conflict and the End of Empire in Egypt*. University of California Press, 1995.

Vitalis, Robert, and Steven Heydemann. "War, Keynesianism, and Colonialism: Explaining State-Market Relations in the Postwar Middle East." In *War, Institutions, and Social Change in the Middle East*. Edited by Steven Heydemann. University of California Press, 2000.

Volait, Mercedes. "Town Planning Schemes for Cairo conceived by Egyptian Planners in the "Liberal Experiment" Period." In *Middle Eastern Cities 1900–1950: Public Places and Public Spheres in Transformation*, edited by Hans Christian Korsholm Nielsen and Jakob Skovgaard-Petersen. Aarhus University Press, 2001.

———. "Making Cairo Modern (1870–1950): Multiple Model for a 'European Style' Urbanism." In *Urbanism: Imported or Exported?*, edited by Joseph Nasr and Mercedes Volait. Wiley-Academy, 2003.

———. "Garden Suburbs, Cairo 1940s–1950s Architect: Mahmoud Riad. Early Experiments with Subsidized Housing." In *Designing Modernity: Architecture in the Arab World 1945–1973*, edited by George Arbid and Philipp Oswalt. JOVIS Verlag GmbH, 2022.

Voskuhl, Adelheid. "Engineering Philosophy: Theories of Technology, German Idealism, and Social Order in High-Industrial Germany." *Technology and Culture* 57, no. 4 (2016): 721–52.

W., W. "Obituary: Baldwin Latham, 1836 to 1917." *Journal of the Royal Sanitary Institute* 38, no. 2 (1917): 99–100.

Wahby, Noura. "Urban Informality and the State: Repairing Cairo's Waters through *Gehood Zateya*." *Environment and Planning E: Nature and Space* 4, no. 3 (2021): 696–717.

Wasfi, Mahmud. *Muhadara 'an Majari al-Qahira*, Jam'iyyat al-Muhandisin al-Malakiyya al-Misriyya. Matba'at al-I'timad, 1939.

Wasfi, Muhammad. *Al-Qawa'id al-Asasiyya fi al-'Imara al-Misriyya*. 2 vols. Madrasat al-Funun wa al-Sana'i' al-Khidiwiyya, 1902.

Waterbury, John. *The Egypt of Nasser and Sadat: The Political Economy of Two Regimes*. Princeton University Press, 1983.

Watson, Hugh. *Sewerage Systems: Their Design and Construction*. Crosby Lockwood and Son, 1911.

Watts, Sheldon. "From Rapid Change to Stasis: Official Responses to Cholera in British-Ruled India and Egypt: 1860 to c. 1921." *Journal of World History* 12, no. 2 (2001): 321–74.

White, Luise. *Speaking with Vampires: Rumor and History in Colonial Africa*. University of California Press, 2000.

Willcocks, William. *Egyptian Irrigation*. 2nd ed. London: E. & F. N. Spon, 1899.

———. *Sixty Years in the East*. William Blackwood and Sons, 1935.

Winch, Donald. "The Problematic Status of the Consumer in Orthodox Economic Thought." In *The Making of the Consumer: Knowledge, Power and Identity in the Modern World*, edited by Frank Trentmann. Berg, 2005.

Wisnioski, Matthew. *Engineers for Change: Competing Visions of Technology in 1960s America*. MIT Press, 2012.

Wittington, Dale, John Waterbury, and Marc Jeuland. "The Grand Renaissance Dam and Prospects for Cooperation on the Eastern Nile." *Water Policy* 16, no. 4 (2014): 595–608.

Wolmar, Christian. *Blood, Iron, and Gold: How the Railways Transformed the World*. Atlantic Books, 2009.

Wright, Arnold, and H. A. Cartwright, eds. *Twentieth Century Impressions of Egypt: Its History, People, Commerce, Industry and Resources*. Lloyd's Greater Britain Publishing Company Ltd., 1909.

Zaki, Ibrahim. *Isti'bad Misr bi-Mashru'at al-Rayy al-Injliziyya*. Al-Matba'a al-Tujariyya al-Haditha, 1928.

INDEX

Page numbers in *italics* refer to figures and tables; those followed by "n" indicate endnotes.

Printed and bound by CPI Group (UK) Ltd, Croydon, CR0 4YY
13/10/2026
14980929-0004